GCSE AQA
Biology
The Revision Guide

This book is for anyone doing **GCSE AQA Biology**.

GCSE Science is all about **understanding how science works**.
And not only that — understanding it well enough to be able to **question** what you hear on TV and read in the papers.

But you can't do that without a fair chunk of **background knowledge**. Hmm, tricky.

Happily this CGP book includes all the **science facts** you need to learn, and shows you how they work in the **real world**. And in true CGP style, we've explained it all as **clearly and concisely** as possible.

It's also got some daft bits in to try and make the whole experience at least vaguely entertaining for you.

What CGP is all about

Our sole aim here at CGP is to produce the highest quality books — carefully written, immaculately presented and dangerously close to being funny.

Then we work our socks off to get them out to you — at the cheapest possible prices.

Contents

How Science Works

Biology 1a — Human Biology

Biology 1b — Evolution and Environment

Biology 2(i) — Life Processes

Biology 2(ii) — Enzymes and Homeostasis

Biology 2(iii) — Genetics

Biology 3(i) — Life Processes 2

Biology 3(ii) — Microorganisms

Exam Skills

Published by Coordination Group Publications Ltd.

From original material by Richard Parsons.

Editors:
Ellen Bowness, Gemma Hallam, Sarah Hilton, Sharon Keeley, Andy Park, Kate Redmond, Alan Rix, Ami Snelling, Claire Thompson, Julie Wakeling.

Contributors:
James Dawson, Sandy Gardner, Adrian Schmit.

ISBN: 978 1 84146 664 4

With thanks to James Foster and Glenn Rogers for the proofreading.
With thanks to Caroline Batten, Laura Phillips and Katie Steele for the copyright research.
With thanks to Forest Research for their permission to reproduce the graphs on page 25.

Graphs to show trend in atmospheric CO_2 *concentration and global temperature on page 5 based on data by EPICA Community Members 2004 and Siegenthaler et al 2005.*
Graph on page 61 — "Pancreatic extracts in the treatment of diabetes mellitus" — Reprinted from, CMAJ March 1922; Page(s) 141-146 by permission of the publisher.

Data used to construct the graph on page 22 provided by the Health Protection Agency.

Groovy website: www.cgpbooks.co.uk

Printed by Elanders Hindson Ltd, Newcastle upon Tyne.
Jolly bits of clipart from CorelDRAW®

Theories Come, Theories Go

SCIENTISTS ARE ALWAYS RIGHT — OR ARE THEY?

Well it'd be nice if that were so, but it just ain't — never has been and never will be.
Increasing scientific knowledge involves making mistakes along the way. Let me explain...

Scientists Come Up with Hypotheses — Then Test Them

1) Scientists try and explain things. Everything.
2) They start by observing or thinking about something they don't understand — it could be anything, e.g. planets in the sky, a person suffering from an illness, what matter is made of... anything.

Hundreds of years ago, we thought demons caused illness.

3) Then, using what they already know (plus a bit of insight), they come up with a hypothesis (a theory) that could explain what they've observed. But a hypothesis is just a theory, a belief. And believing something is true doesn't make it true — not even if you're a scientist.
4) So the next step is to try and convince other scientists that the hypothesis is right — which involves using evidence. First, the hypothesis has to fit the evidence already available — if it doesn't, it'll convince no one.
5) Next, the scientist might use the hypothesis to make a prediction — a crucial step. If the hypothesis predicts something, and then evidence from experiments backs that up, that's pretty convincing. This doesn't mean the hypothesis is true (the 2nd prediction, or the 3rd, 4th or 25th one might turn out to be wrong) — but a hypothesis that correctly predicts something in the future deserves respect.

Other Scientists Will Test the Hypotheses Too

Then we thought it was caused by bad blood (and treated it with leeches).

1) Now then... other scientists will want to use the hypothesis to make their own predictions, and they'll carry out their own experiments. (They'll also try to reproduce earlier results.) And if all the experiments in all the world back up the hypothesis, then scientists start to have a lot of faith in it.
2) However, if a scientist somewhere in the world does an experiment that doesn't fit with the hypothesis (and other scientists can reproduce these results), then the hypothesis is in trouble. When this happens, scientists have to come up with a new hypothesis (maybe a modification of the old theory, or maybe a completely new one).
3) This process of testing a hypothesis to destruction is a vital part of the scientific process. Without the 'healthy scepticism' of scientists everywhere, we'd still believe the first theories that people came up with — like thunder being the belchings of an angered god (or whatever).

If Evidence Supports a Hypothesis, It's Accepted — for Now

1) If pretty much every scientist in the world believes a hypothesis to be true because experiments back it up, then it usually goes in the textbooks for students to learn.

Now we know most illnesses are due to microorganisms.

2) Our currently accepted theories are the ones that have survived this 'trial by evidence' — they've been tested many, many times over the years and survived (while the less good ones have been ditched).
3) However... they never, never become hard and fast, totally indisputable fact. You can never know... it'd only take one odd, totally inexplicable result, and the hypothesising and testing would start all over again.

You expect me to believe that — then show me the evidence...

If scientists think something is true, they need to produce evidence to convince others — it's all part of testing a hypothesis. One hypothesis might survive these tests, while others won't — it's how things progress. And along the way some hypotheses will be disproved — i.e. shown not to be true.
So, you see... not everything scientists say is true. It's how science works.

Your Data's Got to Be Good

Evidence is the key to science — but not all evidence is equally good.
The way evidence is gathered can have a big effect on how trustworthy it is...

Lab Experiments Are Better Than Rumour or Small Samples

1) Results from controlled experiments in laboratories are great. A lab is the easiest place to control variables so that they're all kept constant (except for the one you're investigating). This makes it easier to carry out a fair test. It's also the easiest way for different scientists around the world to carry out the same experiments. (There are things you can't study in a lab though, like climate.)
2) Old wives' tales, rumours, hearsay, "what someone said", and so on, should be taken with a pinch of salt. They'd need to be tested in controlled conditions to be genuinely scientific.
3) Data based on samples that are too small don't have much more credibility than rumours do. A sample should be representative of the whole population (i.e. it should share as many of the various characteristics in the population as possible) — a small sample just can't do that.

Evidence Is Only Reliable If Other People Can Repeat It

Scientific evidence needs to be reliable (or reproducible). If it isn't, then it doesn't really help.

RELIABLE means that the data can be reproduced by others.

In 1989, two scientists claimed that they'd produced 'cold fusion' (the energy source of the Sun — but without the enormous temperatures). It was huge news — if true, this could have meant energy from seawater — the ideal energy solution for the world... forever. However, other scientists just couldn't get the same results — i.e. the results weren't reliable. And until they are, 'cold fusion' isn't going to be generally accepted as fact.

Evidence Also Needs to Be Valid

To answer scientific questions scientists often try to link changes in one variable with changes in another. This is useful evidence, as long as it's valid.

VALID means that the data is reliable AND answers the original question.

EXAMPLE: DO POWER LINES CAUSE CANCER?
Some studies have found that children who live near overhead power lines are more likely to develop cancer. What they'd actually found was a correlation between the variables "presence of power lines" and "incidence of cancer" — they found that as one changed, so did the other. But this evidence is not enough to say that the power lines cause cancer, as other explanations might be possible. For example, power lines are often near busy roads, so the areas tested could contain different levels of pollution from traffic. Also, you need to look at types of neighbourhoods and lifestyles of people living in the tested areas (could diet be a factor... or something else you hadn't thought of...).
So these studies don't show a definite link and so don't answer the original question.

Controlling All the Variables Is Really Hard

In reality, it's very hard to control all the variables that might (just might) be having an effect. You can do things to help — e.g. choose two groups of people (those near power lines and those far away) who are as similar as possible (same mix of ages, same mix of diets etc). But you can't easily rule out every possibility. If you could do a properly controlled lab experiment, that'd be better — but you just can't do it without cloning people and exposing them to things that might cause cancer... hardly ethical.

Does the data really say that?...

If it's so hard to be definite about anything, how does anybody ever get convinced about anything? Well, what usually happens is that you get a load of evidence that all points the same way. If one study can't rule out a particular possibility, then maybe another one can. So you gradually build up a whole body of evidence, and it's this (rather than any single study) that convinces people.

Bias and How to Spot It

Scientific results are often used to help people make a point (e.g. politicians, environmental campaigners... and so on). But results are sometimes presented in a biased way — and you need to be able to spot that.

You Don't Need to Lie to Make Things Biased

1) For something to be misleading, it doesn't have to be untrue. We tend to read scientific facts and assume that they're the 'truth', but there are many different sides to the truth. Look at this headline...

Scientists say 1 in 2 people are of above average weight

Sounds like we're a nation of fatties. It's a scientific analysis of the facts, and almost certainly true.

2) But an average is a kind of 'middle value' of all your data.
Some readings are higher than average (about half of them, usually).
Others will be lower than average (the other half).

So the above headline (which made it sound like we should all lose weight) could just as accurately say:

Scientists say 1 in 2 people are of below average weight

3) The point is... both headlines sound quite worrying, even though they're not. That's the thing... you can easily make something sound really good or really bad — even if it isn't. You can...

① ...use only some of the data, rather than all of it:

"Many people lost weight using the new SlimAway diet. Buy it now!!"

"Many" could mean anything — e.g. 50 out of 5000 (i.e. 1%). But that coud be ignoring most of the data.

② ...phrase things in a 'leading' way:

90% fat free!

Would you buy it if it were "90% cyanide free"? That 10% is the important bit, probably.

③ ...use a statistic that supports your point of view:

The amount of energy wasted is increasing.

Energy wasted per person is decreasing.

The rate at which energy waste is increasing is slowing down.

These describe the same data. But two sound positive and one negative.

Think About Why Things Might Be Biased

1) People who want to make a point can sometimes present data in a biased way to suit their own purposes (sometimes without knowing they're doing it).
2) And there are all sorts of reasons why people might want to do this — for example...
   - Governments might want to persuade voters, other governments, journalists, etc. Evidence might be ignored if it could create political problems, or emphasised if it helps their cause.
   - Companies might want to 'big up' their products. Or make impressive safety claims, maybe.
   - Environmental campaigners might want to persuade people to behave differently.
3) People do it all the time. This is why any scientific evidence has to be looked at carefully. Are there any reasons for thinking the evidence is biased in some way?
   - Does the experimenter (or the person writing about it) stand to gain (or lose) anything? (For example, are they being funded by a particular company or group?)
   - Might someone have ignored some of the data for political or commercial reasons?
   - Is someone using their reputation rather than evidence to help make their case?

Tell me what you want people to believe, and I'll find a statistic to help...

So scientific data and the person presenting it need to be looked at carefully. That doesn't mean the scientific data's always misleading, just that you need to be careful. The most credible argument will be the one that describes all the data that was found, and gives the most balanced view of it.

Science Has Limits

Science can give us amazing things — cures for diseases, space travel, heated toilet seats...
But science has its limitations — there are questions that it just can't answer.

Some Questions Are Unanswered by Science — So Far

1) We don't understand everything. And we never will. We'll find out more, for sure — as more hypotheses are suggested, and more experiments are done. But there'll always be stuff we don't know.

For example, today we don't know as much as we'd like about climate change (global warming). Is climate change definitely happening? And to what extent is it caused by humans?

2) These are complicated questions (see page 5). At the moment scientists don't all agree on the answers. But eventually, we probably will be able to answer these questions once and for all.
3) But by then there'll be loads of new questions to answer.

Other Questions Are Unanswerable by Science

1) Then there's the other type... questions that all the experiments in the world won't help us answer — the "Should we be doing this at all?" type questions. There are always two sides...
2) Take embryo screening (which allows you to choose an embryo with particular characteristics). It's possible to do it — but does that mean we should?
3) Different people have different opinions. For example...

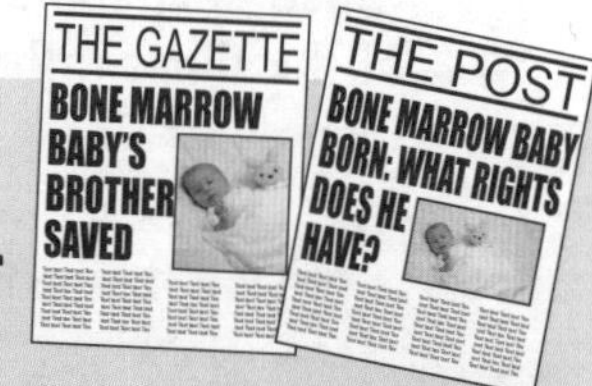

- Some people say it's good... couples whose existing child needs a bone marrow transplant, but who can't find a donor, will be able to have another child selected for its matching bone marrow. This would save the life of their first child — and if they want another child anyway... where's the harm?
- Other people say it's bad... they say it could have serious effects on the child. In the above example the new child might feel unwanted — thinking they were only brought into the world to help someone else. And would they have the right to refuse to donate their bone marrow (as anyone else would)?

4) This question of whether something is morally or ethically right or wrong can't be answered by more experiments — there is no "right" or "wrong" answer.
5) The best we can do is get a consensus from society — a judgement that most people are more or less happy to live by. Science can provide more information to help people make this judgement, and the judgement might change over time. But in the end it's up to people and their conscience.

Loads of Other Factors Can Influence Decisions Too

Here are some other factors that can influence decisions about science, and the way science is used:

Economic factors:
- Companies very often won't pay for research unless there's likely to be a profit in it.
- Society can't always afford to do things scientists recommend without cutting back elsewhere (e.g. investing heavily in alternative energy sources).

Social factors:
- Decisions based on scientific evidence affect people — e.g. should fossil fuels be taxed more highly (to invest in alternative energy)? Should alcohol be banned (to prevent health problems)? Would the effect on people's lifestyles be acceptable...

Environmental factors:
- Genetically modified crops may help us produce more food — but some people say they could cause environmental problems (see page 30).

Science is a "real-world" subject...

Science isn't just done by people in white coats in labs who have no effect on the outside world. Science has a massive effect on the real world every day, and so real-life things like money, morals and how people might react need to be considered. It's why a lot of issues are so difficult to solve.

Climate Change: A Modern Example

So if we're never quite sure of anything, why does anyone bother trying to find stuff out? Well, some scientific stuff is quite important — like avoiding planetary catastrophe, say. Yup, here's global warming...

Taking the Temperature of a Planet Is Hard

Years ago, a French scientist worked out that atmospheric gases, including CO_2, keep the Earth at a temperature that's just right. Later, a Swedish chemist, Arrhenius, predicted that as people burned more coal, the concentration of CO_2 in the atmosphere would rise, and the Earth would get warmer.

1) To test this hypothesis, you need reliable data for two variables — the CO_2 level and temperature. To be valid, the investigation has to cover the whole globe over hundreds of thousands of years (or we'd just discover that it was colder during the last ice age, which we know anyway).
2) To monitor global temperature, scientists often measure the temperature of the sea surface.

The first measurements were done from ships — some bloke would fling a bucket overboard, haul it up and stick a thermometer in it. Later, ships recorded the temperature of the water they took on board to cool their engines. Neither method was exactly great —

- Water samples weren't all taken from the same depth (and deeper water is usually colder).
- The sailors taking the readings were probably a bit slapdash — they were busy sailing.
  So if two samples were taken in the same place, at the same time, the results would quite possibly be different — in other words, not reproducible.
- Ships didn't go everywhere, so the records are a bit patchy. So, you might see that the North Atlantic ocean is getting warmer, but have no idea about the rest of the world. The original hypothesis was about global temperature, so the validity of these results is doubtful.

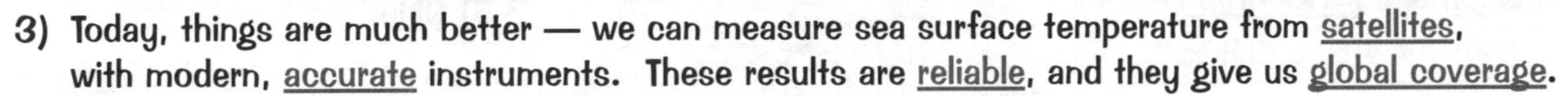

3) Today, things are much better — we can measure sea surface temperature from satellites, with modern, accurate instruments. These results are reliable, and they give us global coverage.
4) We also have very clever ways of finding temperatures and CO_2 levels from the distant past (before thermometers existed) — by examining air bubbles trapped deep in the ice in Antarctica, for example. There are similar tricks involving tree rings, sediments and pollen so the results can be checked. Even so, these methods aren't perfect — there may be contamination problems, for instance.

Interpreting the Data Is Even Harder...

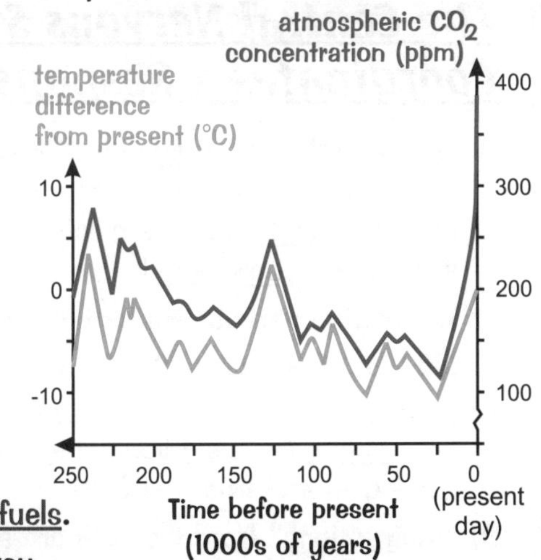

1) This is a graph of the CO_2 and global temperature data. It shows temperature and CO_2 rising very rapidly from about 1850 (when the Industrial Revolution began).
2) But the graph also shows that there have been huge changes in the climate before — you could argue that the recent warming is just part of that natural variability.
3) Scientists now believe that the Earth is warming. There's a scientific consensus that the warming is more than natural variation and that humans are causing it — we're emitting too much CO_2. If that's right, maybe we should stop burning fossil fuels.
4) But there are big interests at stake, and this can influence the way people present the data. If we stopped buying fossil fuels then oil companies, among others, would lose out — so they might emphasise the 'natural variation' argument. People with different interests (like wind turbine manufacturers) might emphasise the more recent rapid rise in temperature and CO_2. They could use exactly the same data — but with a different slant.

Anthropogenic warming — eh?

There's almost no argument about the 'more CO_2 is making us warmer' theory now (there was plenty of doubt previously, when the data was dodgy). And there's a consensus among climate scientists that the warming is 'anthropogenic' (our fault) instead of being down to natural variability in climate. Nice.

The Nervous System

Welcome to AQA Science. First thing on the menu is a page about the nervous system. The nervous system is what lets you react to what goes on around you, so you'd find life tough without it.

Sense Organs Detect Stimuli

A stimulus is a change in your environment which you may need to react to (e.g. a recently pounced tiger). You need to be constantly monitoring what's going on so you can respond if you need to.

1) You have five different sense organs — eyes, ears, nose, tongue and skin.
2) They all contain different receptors. Receptors are groups of cells which are sensitive to a stimulus. They change stimulus energy (e.g. light energy) into electrical impulses.
3) A stimulus can be light, sound, touch, pressure, chemical, or a change in position or temperature.

Sense organs and Receptors
Don't get them mixed up:

The eye is a sense organ — it contains light receptors.

The ear is a sense organ — it contains sound receptors.

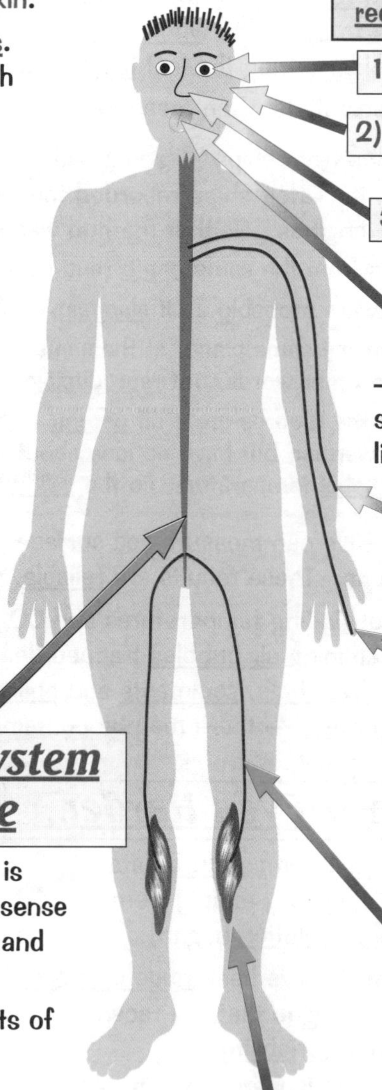

The Five Sense Organs and the receptors that each contains:

1) Eyes — Light receptors.

2) Ears — Sound and "balance" receptors.

3) Nose — Smell receptors — sensitive to chemical stimuli.

4) Tongue — Taste receptors: — sensitive to bitter, salt, sweet and sour, plus the taste of savoury things like monosodium glutamate (MSG) — chemical stimuli.

5) Skin — Sensitive to touch, pressure and temperature change.

Sensory Neurones
The nerve cells that carry signals as electrical impulses from the receptors in the sense organs to the central nervous system.

Motor Neurones
The nerve cells that carry signals to the effector muscles or glands.

Effectors
Muscles and glands are known as effectors — they respond in different ways. Muscles contract in response to a nervous impulse, whereas glands secrete hormones.

The Central Nervous System Coordinates a Response

1) The central nervous system (CNS) is where all the information from the sense organs is sent, and where reflexes and actions are coordinated.

   The central nervous system consists of the brain and spinal cord only.
2) Neurones (nerve cells) transmit the information (as electrical impulses) very quickly to and from the CNS.
3) "Instructions" from the CNS are sent to the effectors (muscles and glands), which respond accordingly.

Your tongue's evolved for Chinese meals — sweet, sour, MSG...

Listen up... the thing with GCSE Science is that it's not just a test of what you know — it's also a test of how well you can apply what you know. For instance, you might have to take what you know about a human and apply it to a horse (easy... sound receptors in its ears, light receptors in its eyes, etc.), or to a snake (so if you're told that certain types of snakes have heat receptors in nostril-like pits on their head, you should be able to work out what type of stimulus those pits are sensitive to). Thinking in an exam... gosh.

Reflexes

Your brain can decide how to respond to a stimulus pretty quickly.
But sometimes, waiting for your brain to make a decision is just too slow. That's why you have reflexes.

Reflexes Help Prevent Injury

1) Reflexes are automatic responses to certain stimuli — they can reduce the chances of being injured.
2) For example, if someone shines a bright light in your eyes, your pupils automatically get smaller so that less light gets into the eye — this stops it getting damaged.
3) Or if you get a shock, your body releases the hormone adrenaline automatically — it doesn't wait for you to decide that you're shocked.
4) The passage of information in a reflex (from receptor to effector) is called a reflex arc.

The Reflex Arc Goes Through the Central Nervous System

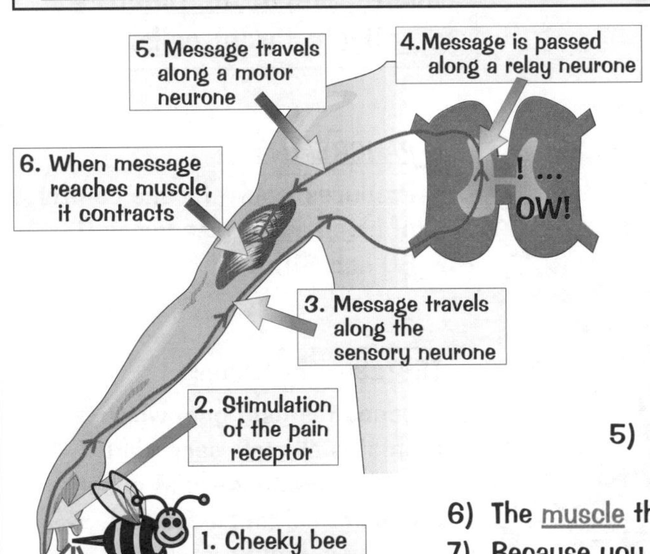

1) The neurones in reflex arcs go through the spinal cord (or an unconscious part of the brain).
2) When a stimulus (e.g. a painful bee sting) is detected by receptors, an impulse is sent along a sensory neurone to the spinal cord.
3) In the spinal cord the sensory neurone passes on the message to another type of neurone — a relay neurone.
4) The relay neurone relays the impulse to a motor neurone.
5) The impulse then travels along the motor neurone to the effector (in this example it's a muscle).
6) The muscle then contracts and moves your hand away from the bee.
7) Because you don't have to think about the response (which takes time) it's quicker than normal responses.

Here's a block diagram of a reflex arc — it shows what happens, from stimulus to response.

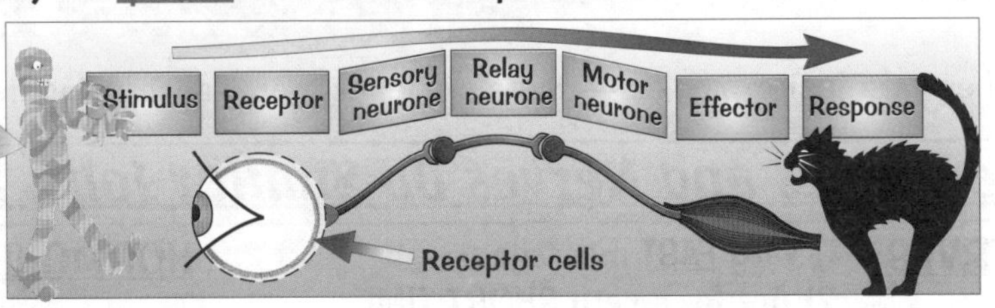

Synapses Connect Neurones

1) The connection between two neurones is called a synapse.
2) The nerve signal is transferred by chemicals which diffuse (move) across the gap.
3) These chemicals then set off a new electrical signal in the next neurone.

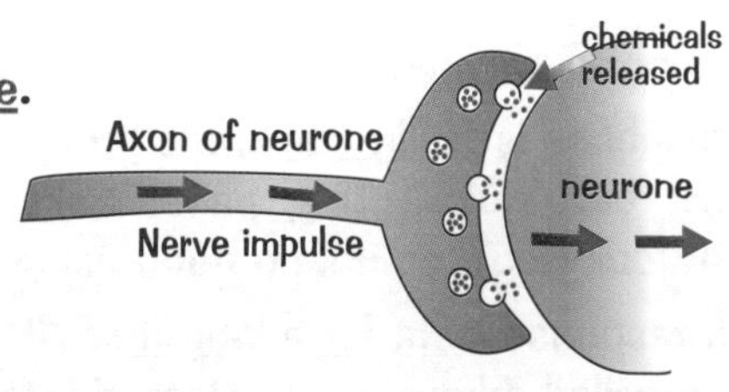

Don't get all twitchy — just learn it...

The difference between a reflex and a "considered response" is the involvement of the conscious part of your brain. Reflexes may bypass your conscious brain completely when a quick response is essential — your body just gets on with things. Reflex actions can be used to assess the condition of unconscious casualties (since the conscious brain isn't involved), or those with spinal injuries (an abnormal reflex could point to where a problem lies). So... if you're asked which bodily system doctors are examining when they tap your knee with a hammer and check that you kick, just work it out. (They're checking parts of your nervous system.)

Hormones

The other way to send information around the body (apart from along nerves) is by using hormones.

Hormones Are Chemical Messengers Sent in the Blood

1) Hormones are chemicals released directly into the blood. They are carried in the blood plasma to other parts of the body, but only affect particular cells (called target cells) in particular places. Hormones control things in organs and cells that need constant adjustment.
2) Hormones are produced in various glands, as shown on the diagram. They travel through your body at "the speed of blood".
3) Hormones tend to have relatively long-lasting effects.

Learn this definition:
HORMONES...
are chemical messengers which travel in the blood to activate target cells.

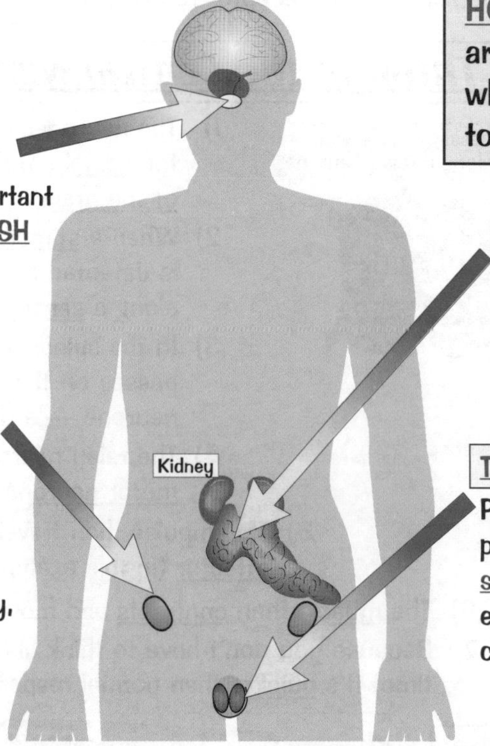

THE PITUITARY GLAND

This produces many important hormones including LH, FSH (see p9) and ADH (which controls water content).

PANCREAS

Produces insulin for the control of blood sugar (see pages 11, 60 and 61).

OVARIES — females only

Produce oestrogen, which controls the menstrual cycle (see p9) and promotes all female secondary sexual characteristics during puberty, e.g. extra body hair and widening of hips.

Kidney

TESTES — males only

Produce testosterone, which promotes all male secondary sexual characteristics at puberty, e.g. extra hair in places and changes in body proportions.

These are just examples — there are loads more, each doing its own thing.

Hormones and Nerves Do Similar Jobs, but There Are Differences

NERVES:
1) Very FAST message.
2) Act for a very SHORT TIME.
3) Act on a very PRECISE AREA.

HORMONES:
1) SLOWER message.
2) Act for a LONG TIME.
3) Act in a more GENERAL way.

So if you're not sure whether a response is nervous or hormonal, have a think...

1) If the response is really quick, it's probably nervous. Some information needs to be passed to effectors really quickly (e.g. pain signals, or information from your eyes telling you about the lion heading your way), so it's no good using hormones to carry the message — they're too slow.
2) But if a response lasts for a long time, it's probably hormonal. For example, when you get a shock, a hormone called adrenaline is released into the body (causing the fight-or-flight response, where your body is hyped up ready for action). You can tell it's a hormonal response (even though it kicks in pretty quickly) because you feel a bit wobbly for a while afterwards.

Nerves, hormones — no wonder revision makes me tense...

Hormones control various organs and cells in the body, though they tend to control things that aren't immediately life-threatening. For example, they take care of all things to do with sexual development, pregnancy, birth, breast-feeding, blood sugar levels, water content... and so on. Pretty amazing really.

The Menstrual Cycle

The monthly release of an egg from a woman's ovaries, and the build-up and breakdown of the protective lining in the womb, is called the menstrual cycle.

The Menstrual Cycle Has Four Stages

Stage 1
Day 1 is when the bleeding starts. The uterus lining breaks down for about four days.

Stage 2
The lining of the womb builds up again, from day 4 to day 14, into a thick spongy layer full of blood vessels, ready to receive a fertilised egg.

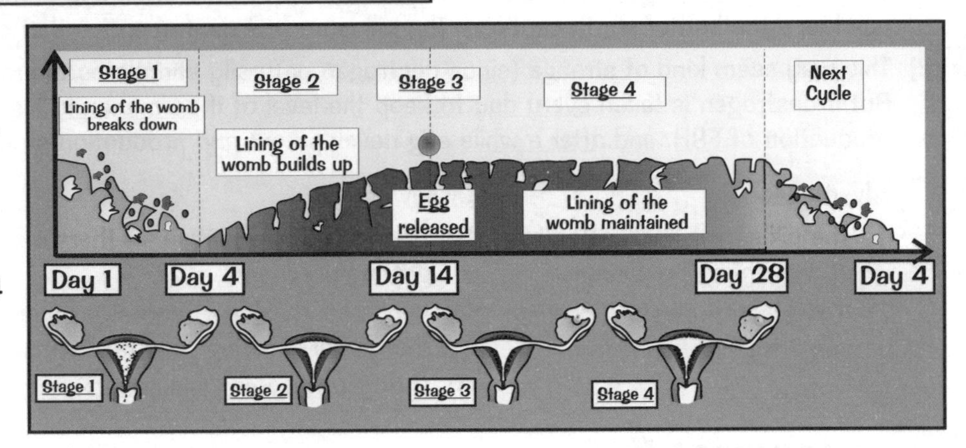

Stage 3 An egg is developed and then released from the ovary at day 14.

Stage 4 The wall is then maintained for about 14 days, until day 28. If no fertilised egg has landed on the uterus wall by day 28, the spongy lining starts to break down again and the whole cycle starts again.

Hormones Control the Different Stages

There are three main hormones involved:

1) FSH (Follicle-Stimulating Hormone):

1) Produced by the pituitary gland.
2) Causes an egg to mature in one of the ovaries.
3) Stimulates the ovaries to produce oestrogen.

2) Oestrogen:

1) Produced in the ovaries.
2) Causes pituitary to produce LH.
3) Inhibits the further release of FSH.

3) LH (Luteinising Hormone):

1) Produced by the pituitary gland.
2) Stimulates the release of an egg at around the middle of the menstrual cycle.

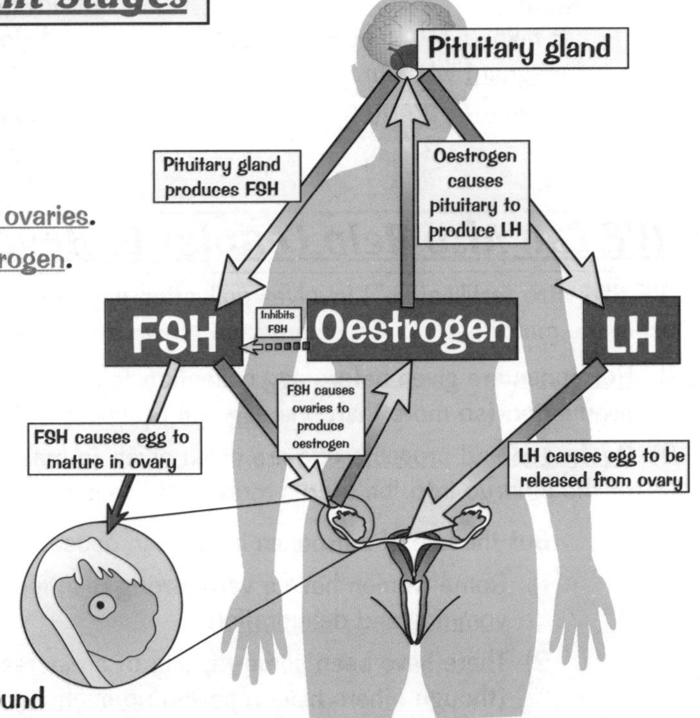

Which came first — the chicken or the luteinising hormone...

In the exam you could be given a completely new situation and have to answer questions about it. For example, say you're told that certain women with epilepsy suffer more seizures at certain points of the menstrual cycle and you have to suggest a reason why. Sounds scary, but the key is not to panic. You know that during the menstrual cycle, hormone levels change — so maybe it's these hormone changes that are triggering the seizures. There are no guarantees, but that'd be a pretty good answer.

Controlling Fertility

The hormones FSH, oestrogen and LH can be used to artificially change how fertile a woman is.

Hormones Can Be Used to Reduce Fertility...

1) The hormone oestrogen can be used to prevent the release of an egg — so oestrogen can be used as a method of contraception. The pill is an oral contraceptive that contains oestrogen.
2) This may seem kind of strange (since oestrogen naturally stimulates the release of eggs). But if oestrogen is taken every day to keep the level of it permanently high, it inhibits the production of FSH, and after a while egg development and production stops and stays stopped.

Advantages
1) The pill's over 99% effective at preventing pregnancy.
2) It reduces the risk of getting some types of cancer.

Disadvantages
1) It isn't 100% effective — there's still a very slight chance of getting pregnant.
2) It can cause side effects like headaches, nausea, irregular menstrual bleeding, and fluid retention.
3) It doesn't protect against sexually transmitted diseases (STDs).

...or Increase It

1) Some women have levels of FSH (Follicle-Stimulating Hormone) that are too low to cause their eggs to mature. This means that no eggs are released and the women can't get pregnant.
2) The hormone FSH can be taken by these women to stimulate egg production in their ovaries. (In fact FSH stimulates the ovaries to produce oestrogen, which stimulates the pituitary gland to produce LH, which stimulates the release of an egg.)

Advantage
It helps a lot of women to get pregnant when previously they couldn't... pretty obvious.

Disadvantages
1) It doesn't always work — some women may have to do it many times, which can be expensive.
2) Too many eggs could be stimulated, resulting in unexpected multiple pregnancies (twins, triplets etc.).

IVF Can Also Help Couples to Have Children

IVF ("in vitro fertilisation") involves collecting eggs from the woman's ovaries and fertilising them in a lab using the man's sperm. These are then grown into embryos, which are transferred to the woman's uterus.

1) Hormones are given before egg collection to stimulate egg production (so more than one egg can be collected).
2) Oestrogen and progesterone are often given to make implantation of the embryo into the uterus more likely to succeed.

But the use of hormones in IVF can cause problems for some women...

1) Some women have a very strong reaction to the hormones — including abdominal pain, vomiting and dehydration.
2) There have been some reports of an increased risk of cancer due to the hormonal treatment (though others have reported no such risk — the position isn't really clear at the moment).

Too many initials to learn — FSH, IVF, STDs, CIA, DVD...

It's not just scientists who have an opinion on whether these hormonal treatments are good or bad... The teachings of several religions are interpreted by some people as being against contraception (though this often includes other methods too — not just hormone-based contraception). Some people also think that contraception increases promiscuous and irresponsible behaviour, since people know they are very unlikely to get pregnant (though they'd still be at risk of being infected by a sexually transmitted disease (STD)). It's one of those situations where science can't really provide all the answers.

Homeostasis

Homeostasis is a fancy word, but it covers lots of things, so maybe that's fair enough.
It means all the functions of your body which try to maintain a "constant internal environment".

Your Body Needs Some Things to Be Kept Constant

All your body's cells are bathed in tissue fluid, which is just blood plasma which has leaked out of the capillaries (on purpose). To keep all your cells working properly, this fluid must be just right — in other words, certain things must be kept at the right level — not too high, and not too low.

Bodily levels that need to be controlled include:

1) Ion content
2) Water content
3) Sugar content
4) Temperature

Ion Content Is Regulated by the Kidneys

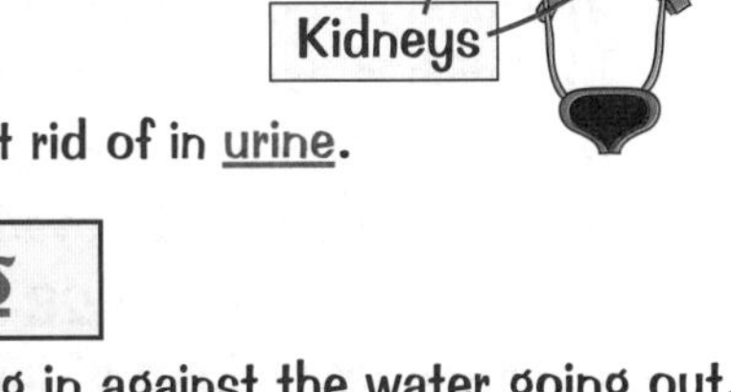

1) Ions (e.g. sodium, Na^+) are taken into the body in food, then absorbed into the blood.
2) If the food contains too much of any kind of ion then the excess ions need to be removed. E.g. a salty meal will contain far too much Na^+.
3) Some ions are lost in sweat (which tastes salty, you'll have noticed).
4) The kidneys will remove the excess from the blood — this is then got rid of in urine.

Water Is Lost from the Body in Various Ways

There's also a need for the body to constantly balance the water coming in against the water going out. Water is taken into the body as food and drink and is lost from the body in these ways:

1) through the SKIN as SWEAT...
2) via the LUNGS in BREATH...
3) via the kidneys as URINE.

Some water is also lost in faeces.

The balance between sweat and urine can depend on what you're doing, or what the weather's like...

On a COLD DAY, or when you're NOT EXERCISING, you don't sweat much, so you'll produce more urine, which will be pale (since the waste carried in the urine is more diluted).

On a HOT DAY, or when you're EXERCISING, you sweat a lot, and so you will produce less urine, but this will be more concentrated (and hence a deeper colour). You will also lose more water through your breath when you exercise because you breathe faster.

Body Temperature Is Controlled by the Brain

1) All enzymes work best at a certain temperature. The enzymes within the human body work best at about 37 °C — and so this is the temperature your body tries to maintain.
2) A part of the brain acts as your own personal thermostat. It's sensitive to the blood temperature in the brain, and it receives messages from the skin that provide information about skin temperature.

Blood Sugar Levels Need to Be Controlled Too

1) Eating foods containing carbohydrate puts glucose (a type of sugar) into the blood from the gut.
2) The normal metabolism of cells removes glucose from the blood.
But if you do a lot of vigorous exercise, then much more glucose is removed.
3) To maintain the right level, you need a way to add or remove glucose to or from the blood — this is the role of the hormone insulin. Diabetes (Type 1) is where your body doesn't produce enough insulin.

My sister never goes out — she's got homeostasis...

Sports drinks (which usually contain electrolytes and carbohydrates) can help your body keep things in order. The electrolytes (e.g. sodium) replace those lost in sweat, while the carbohydrates can give a bit of an energy boost. But claims about sports drinks need to be looked at carefully — see page 19.

Revision Summary 1 for Biology 1a

This is a whopper of a section so here's a page of questions to check that you've understood everything so far. There'd be no point giving you a load of easy-peasy questions, because that's not the sort you'll get in the exam. Have a go at the questions and if you don't know the answer to any of them, then go back and learn the stuff. You should be able to answer most of them by reflex alone, but the last few might need a bit more thinking about.

1) Where would you find the following receptors in a dog: a) smell b) taste c) light d) pressure e) sound?
2) Describe the structure of the central nervous system and explain what it does.
3) What are the following: a) sensory neurone, b) motor neurone, c) effector?
4) What is the purpose of a reflex action?
5) Describe the pathway of a reflex arc from stimulus to response.
6) How do nerve signals get from one neurone to another?
7) Define 'hormone'.
8)* Here's a table of data about response times.
   a) Which response (A or B) is carried by nerves?
   b) Which is carried by hormones?

Response	Reaction time (s)	Response duration (s)
A	0.005	0.05
B	2	10

9) Name one hormone produced by each of the following glands:
   a) pituitary gland, b) ovaries, c) pancreas, d) testes.
10) Draw a timeline of the 28-day menstrual cycle.
    Label the four stages of the cycle and label when the egg is released.
11) What are the two functions of FSH?
12) State two advantages and two disadvantages of using the contraceptive pill.
13) Describe how IVF is carried out.
14)* Water content is kept steady by homeostasis.
    a) Describe how the amount and concentration of urine you produce varies depending on how much exercise you do and how hot it is (assuming you drink the same amount).
    b) The table below shows the amount of water lost in different ways by someone over three days. Match each description with the correct day.

Sitting on Blackpool beach in January

Running a marathon

A relaxing day at home

	Day 1	Day 2	Day 3
Urine	1200 cm^3	1500 cm^3	1000 cm^3
Sweat	1100 cm^3	600 cm^3	3500 cm^3
breath	450 cm^3	450 cm^3	800 cm^3
faeces	100 cm^3	100 cm^3	100 cm^3

    c) How would the amount of water you needed to get from food and water vary between each of these days?
15) Name three other things, besides water content, that need to be kept at the right levels in your body.
16) If you've got too much of a certain ion in your blood, what organ removes it?
17) Why does your body have to maintain a temperature of about 37 °C?

* Answers on page 96

Diet and Exercise

Why is it some people can eat loads and not put on weight, while others only have to look at a chocolate bar and they're a pound heavier...

A Balanced Diet Does a Lot to Keep You Healthy

1) For good health, your diet must provide the energy you need (but not more) — see below.
2) But that's not all. Because the different food groups have different uses in the body, you need to have the right balance of foods as well.
   So you need: ...enough carbohydrates and fats to keep warm and provide energy,
   ...enough protein for growth, cell repair and cell replacement,
   ...enough fibre to keep everything moving smoothly through your digestive system,
   ...and tiny amounts of various vitamins and minerals to keep your skin, bones, blood and everything else generally healthy.

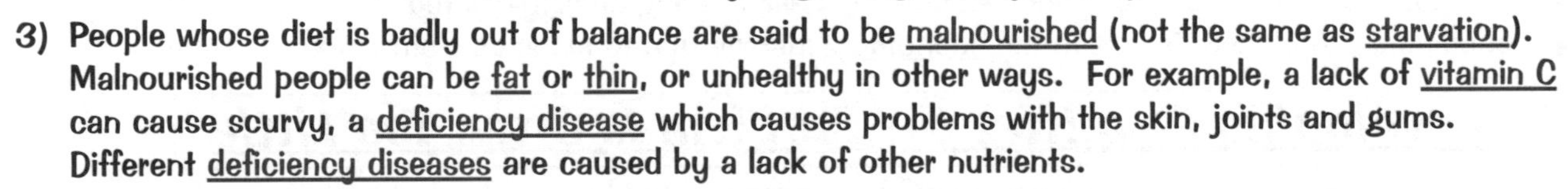

3) People whose diet is badly out of balance are said to be malnourished (not the same as starvation). Malnourished people can be fat or thin, or unhealthy in other ways. For example, a lack of vitamin C can cause scurvy, a deficiency disease which causes problems with the skin, joints and gums. Different deficiency diseases are caused by a lack of other nutrients.

People's Energy Needs Vary Because of Who They Are...

1) You need energy to fuel the chemical reactions in the body that keep you alive. These reactions are called your metabolism, and the speed at which they occur is your metabolic rate.
2) There are slight variations in the resting metabolic rate of different people. For example, muscle needs more energy than fatty tissue, which means (all other things being equal) people with a higher proportion of muscle to fat in their bodies will have a higher metabolic rate.
3) However, physically bigger people are likely to have a higher metabolic rate than smaller people — the bigger you are, the more energy your body needs to be supplied with (because you have more cells).
4) Men tend to have a slightly higher rate than women — they're slightly bigger and have a larger proportion of muscle. Other genetic factors may also have some effect.
5) And regular exercise can boost your resting metabolic rate because it builds muscle.

...and Because of What They Do

1) When you exercise, you obviously need more energy — so your metabolic rate goes up during exercise and stays high for some time after you finish (particularly if the exercise is strenuous).

Activity	kJ/min
Sleeping	4.5
Watching TV	7
Cycling (5 mph)	21
Jogging (5 mph)	40
Climbing stairs	77
Swimming	35
Rowing	58
Slow walking	14

2) So people who have more active jobs need more energy on a daily basis — builders require more energy per day than office workers, for instance. The table shows the average kilojoules burned per minute when doing different activities.
3) The temperature can also affect your metabolic rate. When it's cold, your body has to produce more heat (which requires energy) — this increases your metabolic rate.
4) All these factors have an effect on the amount of energy your diet should contain. If you do little exercise and it's hot outside, you're going to need less energy than if you're constantly on the go in a cold country.

Diet tip — the harder you revise the more calories you burn...

Exercise is important as well as diet — people who exercise regularly are usually fitter than people who don't. But being fit isn't the same as being healthy — e.g. you can be fit as a fiddle and slim, but malnourished at the same time because your diet isn't balanced.

Weight Problems

Health problems due to the wrong kind of diet are different in different parts of the world. In some countries the problem is too much of the wrong kind of food, in others the problem is not having enough.

In Developed Countries the Problem Is Too Much Food

1) In developed countries, obesity is becoming a serious problem. In the UK, 1 in 5 adults are obese, with obesity contributing to the deaths of over 30 000 people each year in England alone.
2) Hormonal problems can lead to obesity, though the usual cause is a bad diet, overeating and a lack of exercise.
3) Health problems that can arise as a result of obesity include: arthritis (inflammation of the joints), diabetes (inability to control blood sugar levels), high blood pressure and heart disease. It's also a risk factor for some kinds of cancer.
4) The National Health Service spends loads each year treating obesity-related conditions. And more is lost to the economy generally due to absence from work.

In Developing Countries the Problem Is Often Too Little

1) In developing countries, some people suffer from lack of food.

2) This can be a lack of one or more specific types of food (malnutrition), or not enough food of any sort (starvation). Young children, the elderly and women tend to be the worst sufferers.
3) The effects of malnutrition vary depending on what foods are missing from the diet. But problems commonly include slow growth (in children), fatigue, poor resistance to infection, and irregular periods in women.

All the Above Claims Are Based on Data

1) If scientists are to decide on the best way to tackle these problems, they need to know as much as possible about them.
2) The first step is to collect some accurate data — but this is very rarely as easy as it sounds. The problem with getting data depends on what data you're trying to get...

Data on Malnutrition

1) People with malnutrition may not reach medical aid. If they do, and if records are being kept detailing causes of death, data can be collated fairly easily (and is usually pretty accurate).
2) However, sometimes medical staff may be dealing with a large-scale emergency, and so they may not have time to keep proper records.

Data on Obesity

1) With obesity the problems are different. The health problems caused by obesity are more long-term, and people don't necessarily seek medical assistance.
2) Surveys can be done and, provided the sample isn't biased, the data might be reliable. But it'll still depend on how you collect the data.
3) For example, not long ago people in the USA were asked to fill in a questionnaire asking about their weight — and obesity was reported in 20% of the population. A later scientific survey based on medical examinations revealed it to be 28% in men and 34% in women.

Obesity is an increasingly weighty issue nowadays...

Whenever you use data, you have to remember that the methods used to collect it are usually far from ideal, which means the information is never perfect. This means the figures are often heavily argued about. (Look at the climate change debate to get a feel for how tricky some of this is.)

Cholesterol and Salt

If you believed everything you read in the papers (and I know I do), you'd think that eating cholesterol and salt are about as sensible as adding cyanide to your chips. But behind the scary stories is a bit of science.

A High Cholesterol Level Is a Risk Factor for Heart Disease

1) Cholesterol is a fatty substance that's essential for good health. It's found in every cell in the body.
2) But you don't want too much of it because a high cholesterol level in the blood causes an increased risk of various problems — like coronary heart disease.
3) This is due to blood vessels getting clogged with fatty cholesterol deposits. This reduces blood flow to the heart, which can lead to angina (chest pain), or a heart attack (if the vessel is blocked completely).
4) The liver is really important in controlling the amount of cholesterol in the body. It makes new cholesterol and removes it from the blood so that it can be eliminated from the body.
5) The amount the liver makes depends on your diet (see below) and inherited factors.

Cholesterol is Carried Around the Body by HDLs and LDLs

1) Cholesterol is transported around the body in the blood by lipoproteins (i.e. fat attached to protein).
2) These can be high density lipoproteins (HDLs), or low density lipoproteins (LDLs).
3) LDLs carry cholesterol from the liver to the body cells — they're sometimes called 'bad cholesterol' as any excess can build up in the arteries. HDLs carry cholesterol that isn't needed from the body cells back to the liver for removal from the body — so they're called 'good cholesterol'.
4) The LDL/HDL balance is very important. Ideally, you want more HDLs than LDLs in the blood.
5) A diet that's low in fat is important (and processed food contains quite a high proportion of fat) — but the types of fat you eat are even more crucial...

SATURATED FATS (with no C=C double bonds) raise cholesterol in the blood by increasing the amount the liver makes and decreasing the amount it gets rid of — so they should be eaten in moderation.

POLYUNSATURATED FATS (with more than one C=C double bond) tend to lower blood cholesterol by increasing its removal from the body and improve the LDL/HDL balance.

MONOUNSATURATED FATS (with exactly one C=C double bond) were long considered to be "neutral" as far as health is concerned. But recent evidence suggests they may also help to lower blood cholesterol and improve the LDL/HDL balance. People who have a diet high in monounsaturates tend to have lower levels of heart disease.

You Need to Watch Your Salt Intake

1) Another risk factor (i.e. something that increases the risk) of heart disease is high blood pressure (hypertension).
2) Eating too much salt may cause hypertension. This is a particular problem for about 30% of the UK population, who are 'salt sensitive' and need to carefully monitor how much salt they eat.

3) However, it's not always easy to keep track of exactly how much salt you eat — most of the salt you eat is probably in processed foods (such as breakfast cereals, soups, sauces, ready meals, biscuits...). The salt you sprinkle on your food makes up quite a small proportion.
4) And as if things weren't complicated enough... on food labels, salt is usually listed as sodium.

My mother-in-law raises my blood pressure...

You need to understand exactly what "risk factor" actually means. If you have a risk factor, it means you're more likely to suffer from a disease, but not that you're guaranteed to. For example, a smoker with high cholesterol and high blood pressure is 30 times more likely to develop heart disease than someone without these risk factors. But it's not a guarantee... statistics don't do guarantees.

Drugs

Drugs alter what goes on in your body. Your body's essentially a seething mass of chemical reactions — drugs can interfere with these reactions, sometimes for the better, sometimes not.

Drugs Change Your Body Chemistry

1) Many drugs are derived from natural substances found in plants, and have been known and used for centuries. For example, heroin is derived from a chemical found in a species of poppy.
2) Some of the chemical changes caused by drugs can lead to the body becoming addicted to the drug. If the drug isn't taken, an addict can suffer physical withdrawal symptoms — and these are sometimes very unpleasant.
3) Heroin and cocaine are very addictive, so are nicotine and caffeine.

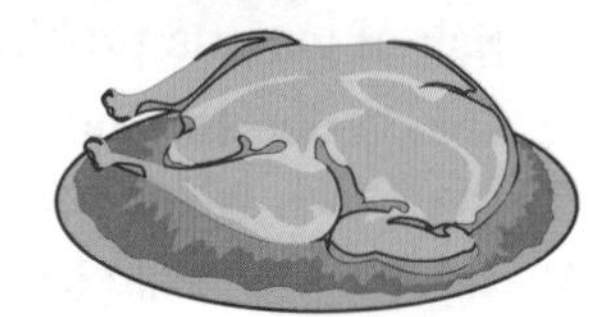

Medical Drugs Have to Be Thoroughly Tested

New drugs are constantly being developed. But before they can be given to the general public, they have to go through a thorough testing procedure. This is what usually happens...

1. Computer models are often used in the early stages — these simulate a human's response to a drug. This can identify promising drugs to be tested in the next stage (but sometimes it's not as accurate as actually seeing the effect on a live organism).
2. Drugs are then developed further by testing on human tissues in the lab. However, you can't use human tissue to test drugs that affect whole or multiple body systems, e.g. testing a drug for blood pressure must be done on a whole animal because it has an intact circulatory system.
3. The next step is to develop and test the drug using live animals. The law in Britain states that any new drug must be tested on two different live mammals. Some people think it's cruel to test on animals, but others believe this is the safest way to make sure a drug isn't dangerous before it's given to humans.
4. After the drug has been tested on animals it's tested on human volunteers in a clinical trial — this should determine whether there are any side effects.

But some people think that animals are so different from humans that testing on animals is pointless.

Things Have Gone Wrong in the Past

An example of what can happen when drugs are not thoroughly tested is the case of thalidomide — a drug developed in the 1950s.

1) Thalidomide was intended as a sleeping pill, and was tested for that use. But later it was also found to be effective in relieving morning sickness in pregnant women.
2) Unfortunately, thalidomide hadn't been tested as a drug for morning sickness, and so it wasn't known that it could pass through the placenta and affect the fetus, causing stunted growth of the fetus's arms and legs. In some cases, babies were born with no arms or legs at all.
3) About 10 000 babies were affected by thalidomide, and only about half of them survived.
4) The drug was banned, and more rigorous testing procedures were introduced.
5) Thalidomide has recently been re-introduced — as a treatment for leprosy, AIDS and certain cancers. But it can't be used on pregnant women.

A little learning is a dangerous thing...

The thalidomide story is an example of an attempt to improve people's lives which then caused some pretty tragic knock-on effects. Could the same thing happen today? Well, maybe not the exact same thing, but there's no such thing as perfect knowledge — we're learning all the time, and you can never eliminate risk completely. Take genetic engineering, for example... those in favour say it could mean no more famines. Those against warn of more unintended consequences. It's a tricky call.

Alcohol and Tobacco

Drugs are also used recreationally. Some of these are legal, others illegal. And some are more harmful than others. But two drugs that have a massive impact on people and society are both legal.

Smoking Tobacco Can Cause Quite a Few Problems

1) Tobacco smoke contains carbon monoxide — this combines irreversibly with haemoglobin in blood cells, meaning the blood can carry less oxygen. In pregnant women, this can deprive the fetus of oxygen, leading to the baby being born underweight.
2) Tobacco smoke also contains carcinogens — chemicals that can lead to cancer. Lung cancer is way more common among smokers than non-smokers (see next page). It's estimated that 90% of lung cancers are associated with smoking (including passive smoking).
3) Disturbingly, the incidence rate (the number of people who get lung cancer) and the mortality rate (the number who die from it) aren't massively different. Put bluntly, this means lung cancer kills most of the people who get it.
4) Smoking also causes disease of the heart and blood vessels (leading to heart attacks and strokes), and damage to the lungs (leading to diseases like emphysema and bronchitis).
5) And the tar in cigarettes damages the cilia (little hairs) in your lungs and windpipe. These hairs, along with mucus, catch a load of dust and bacteria before they reach the lungs. When these cilia are damaged, it's harder for your body to eject stuff that shouldn't be there, which makes chest infections more likely.

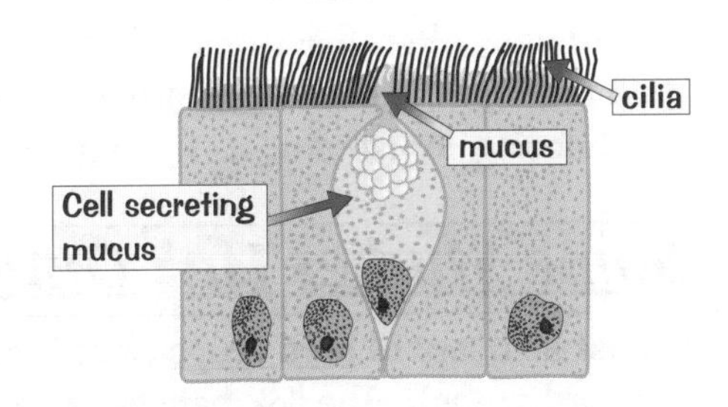

6) And to top it all off, smoking tobacco is addictive — due to the nicotine in tobacco smoke.

Drinking Alcohol Can Do Its Share of Damage Too

1) The main effect of alcohol is to reduce the activity of the nervous system — slowing your reactions.
2) It can also make you feel less inhibited — which can help people to socialise and relax with each other.
3) However, too much leads to impaired judgement, poor balance and coordination, lack of self-control, unconsciousness and even coma.
4) Alcohol in excess also causes dehydration, which can damage brain cells, causing a noticeable drop in brain function. And too much drinking causes severe damage to the liver, leading to liver disease.
5) There are social costs too. Alcohol is linked with way more than half of murders, stabbings and domestic assaults. And alcohol misuse is also a factor in loads of divorces and cases of child abuse.

These Two Legal Drugs Have a Massive Impact

1) Alcohol and tobacco have a bigger impact in the UK than illegal drugs, as so many people take them.
2) The National Health Service spends loads on treating people with lung diseases caused by smoking (or passive smoking). Add to this the cost to businesses of people missing days from work, and the figures get pretty scary.
3) The same goes for alcohol. The costs to the NHS are huge, but are pretty small compared to the costs related to crime (police time, damage to people/property) and the economy (lost working days etc.).
4) And in addition to the financial costs, alcohol and tobacco cause sorrow and anguish to people affected by them, either directly or indirectly.

Drinking and smoking — it's so big and clever...

So it's legal drugs that have the most impact on the country as a whole — when you take everything into consideration. Should the government do more to reduce the number of people who smoke — or is it up to individual people what they do with their lives... there's no easy answer to that one.

Investigating Drugs

It's not always easy to know what to believe when trying to decide about drugs. The best thing to do is to be scientific about it, and look at evidence (though we don't always know as much as we'd like).

Recreational Drugs Can Be Illegal or Legal

1) Illegal drugs are often divided into two main classes — soft and hard. Hard drugs (e.g. heroin and cocaine) are usually thought of as being seriously addictive and generally more harmful.
2) But the terms "soft" and "hard" are a bit vague — they're not scientific descriptions, and you can certainly have problems with soft drug use.

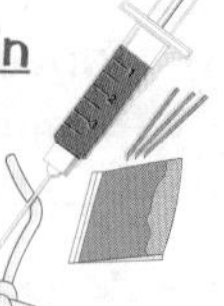

There Are Various Reasons Why People Use Recreational Drugs

So if all these recreational drugs are so dangerous, why do so many people use them...

1) When asked why they use cannabis, most users quote either simple enjoyment, relaxation or stress relief. Some say they do it to get stoned or for inspiration.
2) But very often this turns out to be not the whole story. There may be other factors in the user's background or personal life which influence them in choosing to use drugs. It's a personal thing, and often pretty complicated.

And some multiple sclerosis sufferers say cannabis can relieve pain.

Some Studies Link Cannabis and Hard Drug Use — Others Don't

Almost all users of hard drugs have tried cannabis first (though most users of cannabis do not go on to use hard drugs). The link between cannabis and hard drugs isn't clear, but three opinions are common...

Cannabis is a "stepping stone": The effects of cannabis create a desire to try harder drugs.

Cannabis is a "gateway drug": Cannabis use brings people into contact with drug dealers.

See next page for more info.

It's all down to genetics: Certain people are more likely to take drugs generally, so cannabis users will also try other drugs.

Smoking and Lung Cancer Are Now Known to Be Linked

1) In the first half of the 20th century it was noticed that lung cancer and the popularity of smoking increased together. And studies found that far more smokers than non-smokers got lung cancer.
2) But it was just a statistical correlation at that time (see below) — it didn't prove that smoking caused lung cancer. Some people (especially in the tobacco industry) argued that there was some other factor (e.g. a person's genes) which both caused lung cancer, and also made people more likely to smoke.
3) Later research eventually disproved these claims. Now even the tobacco industry has had to admit that smoking does increase the risk of lung cancer.

Not surprisingly, the "stop-smoking industry" is now big business. The main products available are:

Nicotine gum and patches
These gradually decrease the dose of nicotine (the addictive chemical in tobacco). The success rate with these is about twice that of people using willpower alone.

Acupuncture
People report success with this method, but there is not yet scientific evidence that it works.

Hypnosis
Again, patients report success, but its effects are not scientifically proven.

"I was doing research, man" — tell that to the judge...

If two quantities are statistically correlated, it means they rise or fall together, or one falls as the other rises. But this doesn't automatically mean that a change in one causes the change in the other.
For example, a study showed that villages with more storks nesting in them had higher birth rates (true) — but that doesn't mean the storks were causing babies to be born.
(The number of storks and the number of births both depended on another quantity — the number of people in the village. The more people there were, the more babies were born, and the more houses (and chimneys) there were for storks to nest in.)

Health Claims

It's sometimes hard to figure out if health claims or adverts are true or not.

New Day, New Food Claim — It Can't All Be True

1) To get you to buy a product, advertisers aren't allowed to make claims that are untrue — that's illegal.
2) But they do sometimes make claims that could be misleading or difficult to prove (or disprove).

   For example, some claims are just vague (calling a product "light" for instance — does that mean low calorie, low fat, something else...).

   Alternatively, they might call a breakfast cereal "low fat", and that'd be true.
   But that could suggest that other breakfast cereals are high in fat — when in fact they're not.
3) And every day there's a new food scare in the papers (eeek — we're all doomed).
   Or a new miracle food (phew — we're all saved).
4) It's not easy to decide what to believe and what to ignore. But these things are worth looking for:

   a) Is the report a scientific study, published in a reputable journal?
   b) Was it written by a qualified person (not connected with the food producers)?
   c) Was the sample of people asked/tested large enough to give reliable results?
   d) Have there been other studies which found similar results?

A "yes" to one or more of these is a good sign.

Not All Diets Are Scientifically Proven

With each new day comes a new celebrity-endorsed diet. It's a wonder anyone's overweight.

1) A common way to promote a new diet is to say, "Celebrity A has lost x pounds using it".
2) But effectiveness in one person doesn't mean much. Only a large survey can tell if a diet is more or less effective than just eating less and exercising more — and these aren't done often.
3) The Atkins diet was high profile, and controversial — so it got investigated.
   People on the diet certainly lost weight. But the diet's effect on general health (especially long-term health) has been questioned. The jury's still out.
4) Weight loss is a complex process. But just like with food claims, the best thing to do is look at the evidence in a scientific way.

It's the Same When You Look at Claims About Drugs

Claims about the effects of drugs (both medical and illegal ones) also need to be looked at critically.
But at least here the evidence is usually based on scientific research.

STATINS

1) There's evidence that drugs called statins lower blood cholesterol and significantly lower the risk of heart disease in diabetic patients.
2) The original research was done by government scientists with no connection to the manufacturers. And the sample was big — 6000 patients.
3) It compared two groups of patients — those who had taken statins and those who hadn't. Other studies have since backed up these findings.

So control groups were used. And the results were reproducible.

But research findings are not always so clear cut...

CANNABIS

1) Many scientists have looked at whether cannabis use causes brain damage and mental health problems or leads to further drug taking. The results vary, and are sometimes open to different interpretations.
2) Basically, until more definite scientific evidence is found, no one's sure.

"Brad Pitt says it's great" is NOT scientific proof...

Learn what to look out for before you put too much faith in what you read. Then buy my book — 100% of the people I surveyed (i.e. both of them) said it had no negative affect whatsoever on their overall well-being!

Fighting Disease

When certain microorganisms (called pathogens) enter the body they cause disease.

There Are Two Main Types of Pathogen: Bacteria and Viruses

...and they can both multiply quickly inside your body — they love the warm conditions.

1. Bacteria Are Very Small Living Cells

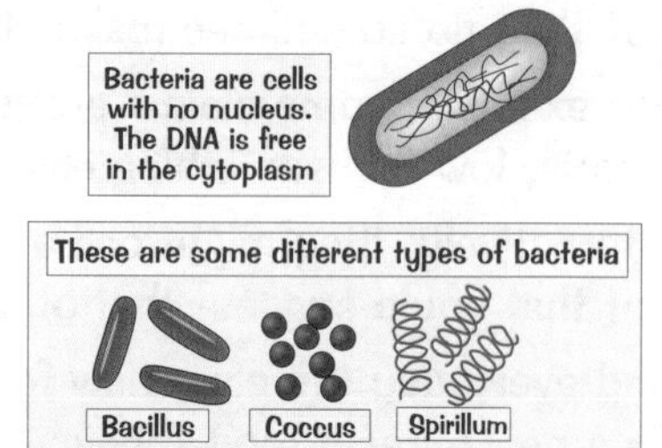

1) Bacteria are very small cells (about 1/100th the size of your body cells), which can reproduce rapidly inside your body.
2) They make you feel ill by doing two things:
   a) damaging your cells, b) producing toxins (poisons).
3) But... some bacteria are useful if they're in the right place, like in your digestive system.

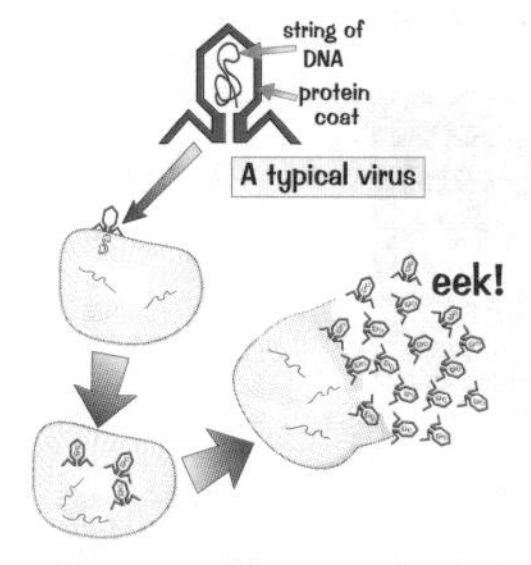

A typical virus

2. Viruses Are Not Cells — They're Much Smaller

1) Viruses are not cells. They're tiny, about 1/100th the size of a bacterium. They're usually no more than a coat of protein around some genetic material.
2) They replicate themselves by invading your cells and using the cells' machinery to produce many copies of themselves. The cell will usually then burst, releasing all the new viruses.
3) This cell damage is what makes you feel ill.

Your Body Has a Pretty Sophisticated Defence System

1) Your skin, plus hairs and mucus in your respiratory tract (breathing pipework), stop a lot of nasties getting inside your body.
2) And to try and prevent microorganisms getting into the body through cuts, small fragments of cells (called platelets) help blood clot quickly to seal wounds. If the blood contains low numbers of platelets then it will clot more slowly.
3) But if something does make it through, your immune system kicks in. The most important part of your immune system is the white blood cells. They travel around in your blood and crawl into every part of you, constantly patrolling for microbes. When they come across an invading microbe they have three lines of attack.

1. Consuming Them

White blood cells can engulf foreign cells and digest them.

2. Producing Antibodies

1) Every invading cell has unique molecules (called antigens) on its surface.
2) When your white blood cells come across a foreign antigen (i.e. one it doesn't recognise), they will start to produce proteins called antibodies to lock on to and kill the invading cells. The antibodies produced are specific to that type of antigen — they won't lock on to any others.

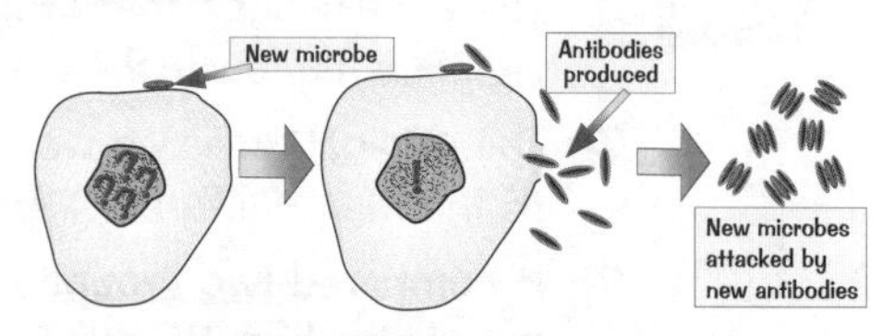

3) Antibodies are then produced rapidly and flow all round the body to kill all similar bacteria or viruses.
4) If the person is infected with the same pathogen again the white blood cells will rapidly produce the antibodies to kill it — the person is naturally immune to that pathogen and won't get ill.

3. Producing Antitoxins

These counter toxins produced by the invading bacteria.

Fight disease — blow your nose with boxing gloves...

So by now you might have worked out that if you have a low level of white blood cells you'll be more susceptible to infections. In fact, HIV/AIDS doesn't kill people directly, it just makes it easier for something else to by attacking white blood cells and weakening the immune system. However, other diseases (e.g. leukaemia) can increase the number of white blood cells — and that's no good either.

Fighting Disease

Dealing with disease comes in three basic flavours — prevention, dealing with symptoms, and curing.

Immunisation — Protects from Future Infections

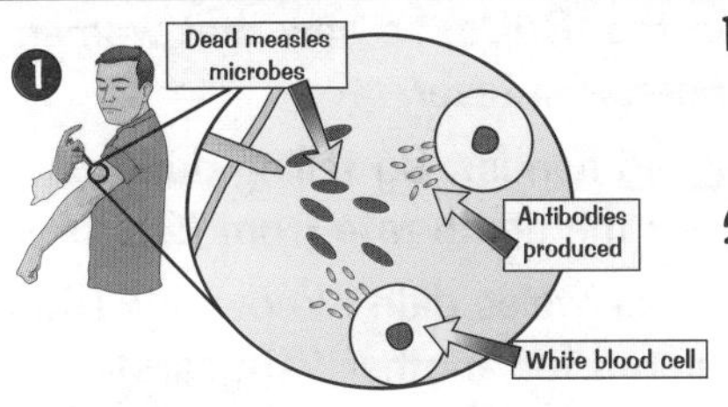

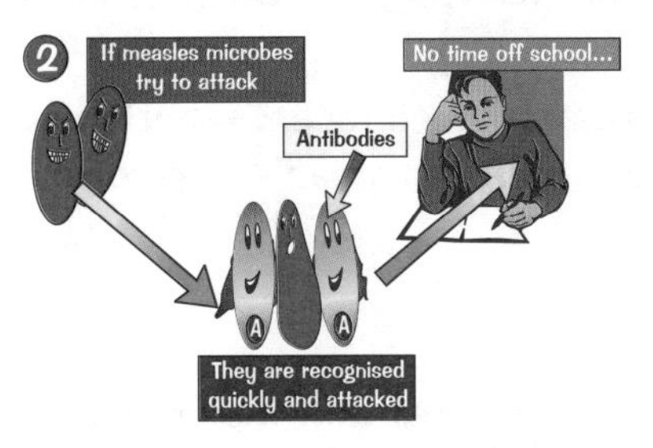

1) When you're infected with a new microorganism, it takes your white blood cells a few days to learn how to deal with it. But by that time, you can be pretty ill.
2) Immunisation involves injecting dead or inactive microorganisms. These carry antigens, which cause your body to produce antibodies to attack them — even though the microorganism is harmless (since it's dead or inactive).
   For example, the MMR vaccine contains weakened versions of the viruses that cause measles, mumps and rubella (German measles) stuck together.
3) But if live microorganisms of the same type appear after that, the white blood cells can rapidly mass-produce antibodies to kill off the pathogen. Cool.
4) Vaccinations "wear off" over time. So booster injections can be given to increase levels of antibodies again.

Some Drugs Just Relieve Symptoms — Others Cure the Problem

A horrid Flu Virus

1) Painkillers (e.g. aspirin) are drugs that relieve pain (no, really). However, they don't actually tackle the cause of the disease, they just help to reduce the symptoms.
2) Other drugs do a similar kind of thing — reduce the symptoms without tackling the underlying cause. For example, lots of "cold remedies" don't actually cure colds.
3) Antibiotics (e.g. penicillin) work differently — they actually kill (or harm) the bacteria causing the problem without killing your own body cells.
4) However, antibiotics don't destroy viruses. Viruses reproduce using your own body cells which makes it very difficult to develop drugs that destroy just the virus without killing the body's cells.
5) Flu and colds are caused by viruses. Usually you just have to wait for your body to deal with the virus, and relieve the symptoms if you start to feel really grotty. There are some antiviral drugs available, but they're usually reserved for very serious viral illnesses (such as AIDS and hepatitis).

Bacteria Can Become Resistant to Antibiotics

1) Antibiotics were an incredibly important (but accidental) discovery.
   Some killer diseases (e.g. pneumonia and tuberculosis) suddenly became much easier to treat.
   The 1940s are sometimes called the era of the antibiotics revolution — it was that big a deal.
2) Unfortunately, bacteria evolve (adapt to their environment). If antibiotics are taken to deal with an infection but not all the bacteria are killed, those that survive may be resistant to the antibiotic and go on to flourish. This process (an example of natural selection) leaves you with an antibiotic-resistant strain of bacteria — not ideal.
3) A good example of antibiotic-resistant bacteria is MRSA (methicillin-resistant *Staphylococcus aureus*) — it's resistant to the powerful antibiotic methicillin.
4) This is why it's important for patients to always finish a course of antibiotics, and for doctors to avoid over-prescribing them.

Antibiotic resistance is inevitable...

Antibiotic resistance is scary. Bacteria reproduce quickly, and so are pretty fast at evolving to deal with threats (e.g. antibiotics). If we were back in the situation where we had no way to treat bacterial infections, we'd have a nightmare. So do your bit, and finish your courses of antibiotics.

Treating Disease — Past and Future

The treatment of disease has changed somewhat over the last 200 years or so.

Semmelweiss Cut Deaths by Using Antiseptics

1) While Ignaz Semmelweiss was working in Vienna General Hospital in the 1840s, he saw that women were dying in huge numbers after childbirth from a disease called puerperal fever.
2) He believed that doctors were spreading the disease on their unwashed hands. By telling doctors entering his ward to wash their hands in an antiseptic solution, he cut the death rate from 12% to 2%.
3) The antiseptic solution killed bacteria on doctors' hands, though Semmelweiss didn't know this (the existence of bacteria and their part in causing disease wasn't discovered for another 20 years). So Semmelweiss couldn't prove why his idea worked, and his methods were dropped when he left the hospital (allowing death rates to rise once again — d'oh).
4) Nowadays we know that basic hygiene is essential in controlling disease (though recent reports have found that a lack of it in some modern hospitals has helped the disease MRSA spread — see below).

Vaccinations Also Help to Prevent Disease

Vaccinations have changed the way we fight disease. We don't always have to deal with the problem once it's happened — we can prevent it happening in the first place. (See page 21 for more on vaccines.)

1) Vaccines have helped control lots of infectious diseases that were once common in the UK (e.g. polio, measles, whooping cough, rubella, mumps, tetanus...).
2) And if an outbreak does occur, vaccines can slow down or stop the spread (if people don't catch the disease, they won't pass it on).
3) Vaccination is now used all over the world. Smallpox no longer occurs at all, and polio infections have fallen by 99%.
4) Vaccines don't always work — sometimes they don't give you immunity, and sometimes you can have a bad reaction (e.g. swelling, or maybe something more serious such as a fever or seizures). But bad reactions are very rare. Vaccines are very safe (see below).

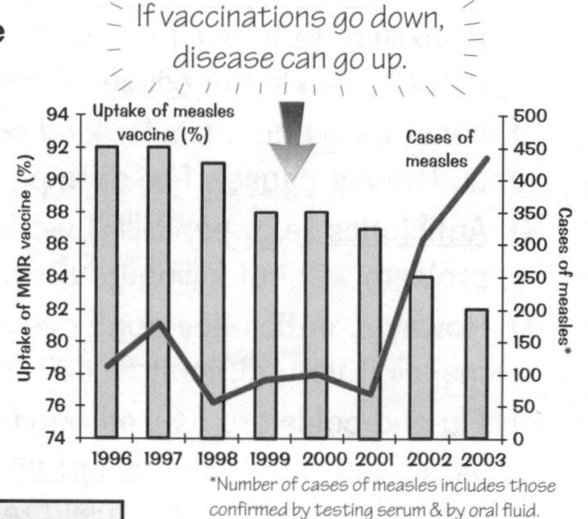

*Number of cases of measles includes those confirmed by testing serum & by oral fluid.

We Face New and Scary Dangers All the Time

1) For the last few decades, humans have been able to deal with bacterial infections pretty easily using antibiotics.
2) But bacteria evolve — MRSA bacteria are already resistant to certain antibiotics (see page 21).
3) And there'd be a real problem if a virus evolved so that it was both deadly and could easily pass from person to person. (Flu viruses, for example, evolve quickly so this is quite possible.)
4) If this happened, precautions could be taken to stop the virus spreading in the first place (though this is hard nowadays — millions of people travel by plane every day). And vaccines and antiviral drugs could be developed (though these take time to mass produce).
5) But in the worst-case scenario, a flu pandemic (e.g. one evolved from bird flu) could kill billions of people all over the world.

A pandemic is when a disease spreads all over the world.

Prevention is better than cure...

Deciding whether to have a vaccination means balancing risks — the risk of catching the disease if you don't have a vaccine, against the risk of having a bad reaction if you do. As always, you need to look at the evidence. For example, if you get measles (the disease), there's about a 1 in 15 chance that you'll get complications (e.g. pneumonia) — and about 1 in 500 people who get measles actually die. However, the number of people who have a problem with the vaccine is more like 1 in 1 000 000.

Revision Summary 2 for Biology 1a

Congratulations, you've made it to the end of the first section. I reckon that section wasn't too bad, there's some pretty interesting stuff there — drugs, booze, diets, vaccinations... what more could you want? Actually, I know what more you could want — some questions to make sure you know it all.

1) Name the six food groups that you need in a balanced diet.
2)* Put these people in order of how much energy they are likely to need from their food (from highest to lowest): a) builder, b) professional runner, c) waitress, d) secretary.
3) Name five health problems that are associated with obesity.
4) Explain why it may be difficult to get accurate data on: a) malnutrition, b) obesity.
5) Explain what is meant by 'good cholesterol' and 'bad cholesterol'.
6) Why is it dangerous to have high levels of cholesterol?
7) Why is it bad to eat too much salt?
8) If you never sprinkle any salt onto your food, are you safe from having too much salt in your diet?
9) Describe the four stages of testing drugs for medicinal use.
10) Name a drug that was not tested thoroughly enough and describe the consequences of its use.
11) How do carbon monoxide, carcinogens and tar in tobacco smoke each affect the body?
12)* Here is a graph of Mark's blood alcohol concentration against time.
   a) When did Mark have his first alcoholic drink?
   b) When did Mark have his second alcoholic drink?
   c) The legal limit for driving is 80 mg of alcohol per 100 ml of blood. Would Mark have been legally allowed to drive at 9 pm?

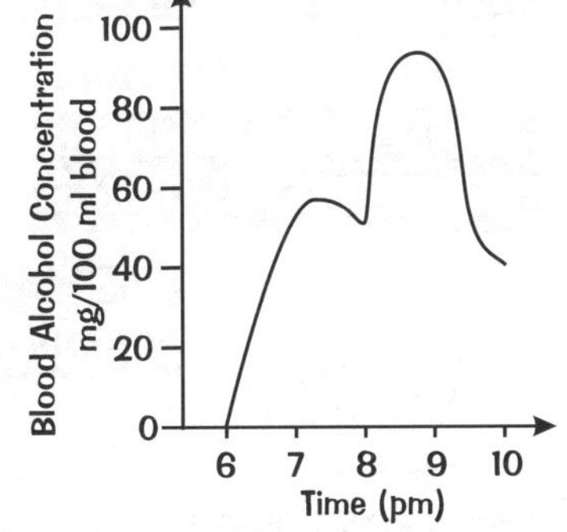

13) Explain three opinions about the link between cannabis and hard drug use.

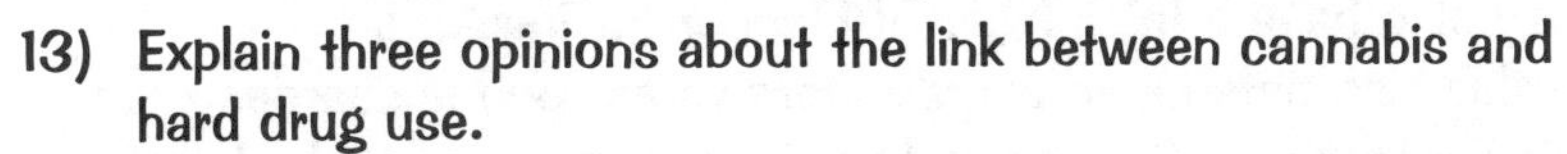

14) Describe three ways that could help someone stop smoking. Are all of these ways proven to work?
15) Name four things that you could consider when trying to decide if a health claim is believable.
16)* A new sort of diet bread is brought out called Baka-Lite. It proudly claims to have only 48 calories (kcal) per slice.
   a) Compare its nutritional information to that of Standard Bread and comment on its claim.
   b) Overall, which bread do you think is the healthier choice?

Standard Bread — NUTRITIONAL INFORMATION

	Per slice	Per 100 g
Energy	92 kcal	184 kcal
Protein	4.6 g	9.2 g
Carbohydrates	20.3 g	40.5 g
Fat (of which saturates)	2.3 g (0.4 g)	4.5 g (0.7 g)
Fibre	1.6 g	3.1 g
Sodium (salt equivalent)	0.22 g (0.57 g)	0.44 g (1.13 g)

Baka-Lite Bread — NUTRITIONAL INFORMATION

	Per slice	Per 100 g
Energy	48 kcal	192 kcal
Protein	2.2 g	8.8 g
Carbohydrates	11.2 g	44.8 g
Fat (of which saturates)	1.8 g (0.3 g)	7.2 g (1.2 g)
Fibre	0.2 g	0.8 g
Sodium (salt equivalent)	0.22 g (0.57 g)	0.88 g (2.26 g)

17) Explain the difference between bacteria and viruses.
18) Describe the three ways in which your immune system defends the body against disease.
19) Explain how immunisation can stop someone getting ill.
20) What problem can be made worse by the over-prescription of antibiotics?
21) What practice did Semmelweiss introduce in the 1840s. Explain why this reduced death rates on his ward.

* Answers on page 96

Adapt and Survive

Animals and plants survive in many different environments — from hot deserts to cold polar regions, and pretty much everywhere in between — they can do this because they have adapted to their environment.

Desert Animals Have Adapted to Save Water

Animals that live in hot, dry conditions need to keep cool and use water efficiently.

LARGE SURFACE AREA COMPARED TO VOLUME

This lets desert animals lose more body heat — which helps to stop them overheating.

EFFICIENT WITH WATER

1) Desert animals lose less water by producing small amounts of concentrated urine.
2) They also make very little sweat. Camels are able to do this by tolerating big changes in body temperature, while kangaroo rats live in burrows underground where it's cool.

GOOD IN HOT, SANDY CONDITIONS

1) Desert animals have very thin layers of body fat to help them lose body heat. Camels keep nearly all their fat in their humps.
2) Large feet spread their weight across soft sand — making getting about easier.
3) A sandy colour gives good camouflage — so they're not as easy for their predators to spot.

Arctic Animals Have Adapted to Reduce Heat Loss

Animals that live in really cold conditions need to keep warm.

SMALL SURFACE AREA COMPARED TO VOLUME

Animals living in cold conditions have a compact (rounded) shape to keep their surface area to a minimum — this reduces heat loss.

WELL INSULATED

1) They also have a thick layer of blubber for insulation — this also acts as an energy store when food is scarce.
2) Thick hairy coats keep body heat in, and greasy fur sheds water (this prevents cooling due to evaporation).

GOOD IN SNOWY CONDITIONS

1) Arctic animals have white fur to match their surroundings — for camouflage.
2) Big feet help by spreading weight — which stops animals sinking into the snow or breaking thin ice.

Some Plants Have Adapted to Living in a Desert

Desert-dwelling plants make best use of what little water is available.

MINIMISING WATER LOSS

1) Cacti have spines instead of leaves — to reduce water loss.
2) They also have a small surface area compared to their size (about 1000 times smaller than normal plants), which also reduces water loss.
3) A cactus stores water in its thick stem.

MAXIMISING WATER ABSORPTION

Some cacti have shallow but extensive roots to absorb water quickly over a large area. Others have deep roots to access underground water.

Some Plants and Animals Are Adapted to Deter Predators

There are various special features used by animals and plants to help protect them against being eaten.

1) Some plants and animals have armour — like roses (with thorns), cacti (with sharp spines) and tortoises (with hard shells).
2) Others produce poisons — like bees and poison ivy.
3) And some have amazing warning colours to scare off predators — like wasps.

In a nutshell, it's horses for courses...

It's no accident that animals and plants look like they do. So by looking at an animal's characteristics, you should be able to have a pretty good guess at the kind of environment it lives in — or vice versa. Why does it have a large/small surface area... what are those spines for... why is it green... and so on.

Populations and Competition

Often different organisms have adapted to suit the same environment and use the same resources — which means that sooner or later they'll probably end up competing with each other for survival...

The Size of Any Population Depends on Three Main Factors

A population is a group of organisms of one species that live in a particular environment.
The size of a population will go up and down due to THREE MAIN FACTORS:

1) COMPETITION

Organisms compete with other species (and members of their own species) for the same resources.

Plants and animals compete in similar ways:

a) Plants often compete with each other for light, water and nutrients from the soil.

b) Animals often compete with each other for space (territory), food, water and mates — e.g. red and grey squirrels live in the same habitat and eat the same food. Competition with the grey squirrels for these resources means there's not enough food for the reds — so the population of red squirrels is decreasing (see below).

2) DISEASE

Infectious diseases caused by bacteria and viruses can kill off many members of a population — but organisms that are fit and healthy stand the best chance of survival.

3) PREDATION

If an organism gets eaten its population will decrease. For example, humans have eaten a lot of cod, which has reduced the cod population dramatically.

Organisms Compete for Resources

Competition for resources can affect the location, size and distribution of populations in the wild.

1) Organisms will live where they can find the resources they need to survive (food, water, etc.). For example, a puffin eats small seafish, so it lives by the sea.
2) Competition for the same resources means that a habitat will only be able to support a certain number of organisms. If the amount of resources in an area decreases, the size of a population there will also decrease — because the organisms will either die or move to where there are more resources.
3) Competition also affects how far apart members of a population are in a habitat (i.e. the distribution of members of the population). If there aren't many resources an organism will need a lot of space to find enough food, water etc. — but if there are loads of resources you can have loads of organisms in a smaller space.

e.g. Red and Grey Squirrels

In 1876 the grey squirrel was introduced into the UK. The native red squirrels were unable to compete very well with the larger grey squirrels — causing the red squirrel population to decrease.

The grey squirrel is better adapted to deciduous woodland than the red squirrel and so outcompetes it. Red squirrels can only outcompete grey squirrels in coniferous woodland because they can eat conifer seeds, so their distribution is now mostly confined to pine forests.

Distribution of red squirrels, 1998

Distribution of grey squirrels, 1998

maps courtesy of Forest Research

I compete with my brother for the front seat of the car...

In the exam you might get asked about the distribution of any animals or plants. Just think about what the organisms would need to survive. And remember, if things are in limited supply then there's going to be competition. And the more similar the needs of the organisms, the more they'll have to compete.

Variation in Plants and Animals

You'll probably have noticed that not all people are identical. There are reasons for this.

Organisms of the Same Species Have Differences

1) Different species look... well... different — my dog definitely doesn't look like a daisy.
2) But even organisms of the same species will usually look at least slightly different — e.g. in a room full of people you'll see different colour hair, individually shaped noses, a variety of heights etc.
3) These differences are called the variation within a species — and there are two types of variation: genetic variation and environmental variation.

Different Genes Cause Genetic Variation

1) All plants and animals have characteristics that are in some ways similar to their parents' (e.g. I've got my dad's nose, apparently).
2) This is because an organism's characteristics are determined by the genes inherited from their parents. (Genes are the codes inside your cells that control how you're made — more about these on p27.)
3) Most animals (and quite a lot of plants) get some genes from the mother and some from the father.
4) This combining of genes from two parents causes genetic variation — no two of the species are genetically identical (other than identical twins).
5) Some characteristics are determined only by genes (e.g. a plant's flower colour). In animals these include: eye colour, blood group and inherited disorders (e.g. haemophilia or cystic fibrosis).

Characteristics are also Influenced by the Environment

1) The environment that organisms live and grow in also causes differences between members of the same species — this is called environmental variation.
2) Environmental variation covers a wide range of differences — from losing your toes in a piranha attack, to getting a suntan, to having yellow leaves (never happened to me yet though), and so on.
3) Basically, any difference that has been caused by the conditions something lives in, is an environmental variation.

A plant grown on a nice sunny windowsill would grow luscious and green.

The same plant grown in darkness would grow tall and spindly and its leaves would turn yellow — these are environmental variations.

Most Characteristics are Due to Genes AND the Environment

1) Most characteristics (e.g. body weight, height, skin colour, condition of teeth, academic or athletic prowess, etc.) are determined by a mixture of genetic and environmental factors.
2) For example, the maximum height that an animal or plant could grow to is determined by its genes. But whether it actually grows that tall depends on its environment (e.g. how much food it gets).

My mum's got no trousers — cos I've got her jeans...

So, you are the way you are partly because of the genes you inherited off your folks. But you can't blame it all on your parents, since your environment then takes over and begins to mould you in all sorts of ways. In fact, it's often really tricky to decide which factor is more influential, your genes or the environment — a good way to study this is with identical twins.

Genes, Chromosomes and DNA

This page is a bit tricky, but it's dead important you get to grips with all the stuff on it — because you're going to hear a lot more about it over the next few pages...

1) Most cells in your body have a nucleus — and it's the nucleus that contains your genetic material.

nucleus

2) The human cell nucleus contains 23 pairs of chromosomes. They are all well known and numbered. We all have two No. 19 chromosomes and two No. 12s etc.

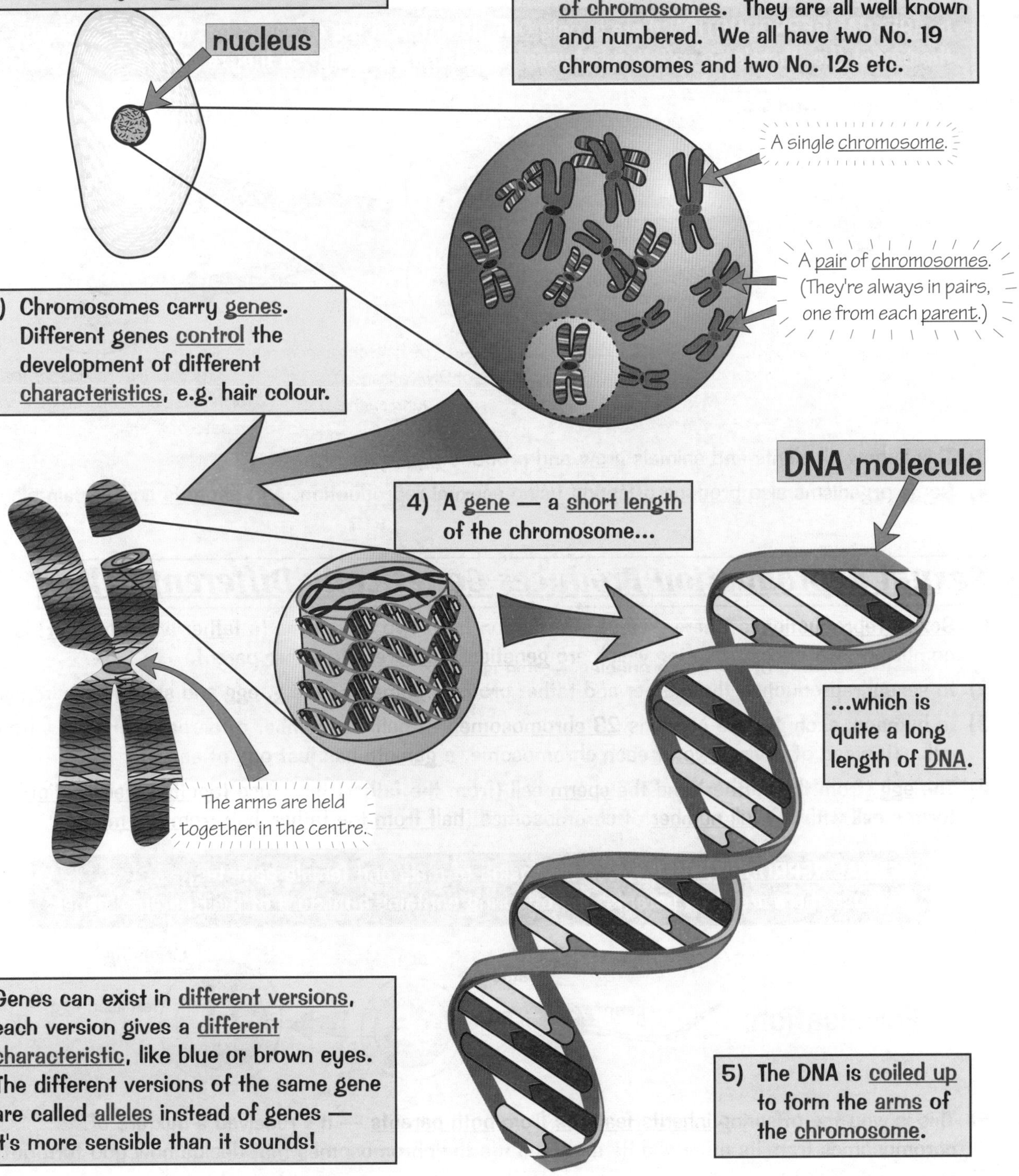

A single chromosome.

A pair of chromosomes. (They're always in pairs, one from each parent.)

3) Chromosomes carry genes. Different genes control the development of different characteristics, e.g. hair colour.

DNA molecule

4) A gene — a short length of the chromosome...

...which is quite a long length of DNA.

The arms are held together in the centre.

Genes can exist in different versions, each version gives a different characteristic, like blue or brown eyes. The different versions of the same gene are called alleles instead of genes — it's more sensible than it sounds!

5) The DNA is coiled up to form the arms of the chromosome.

It's hard being a DNA molecule, there's so much to remember...

This is the nitty gritty of genetics, so you definitely need to understand everything on this page or you'll find the rest of this topic dead hard. The best way to get all of these important facts engraved in your mind is to cover the page, scribble down the main points and sketch out the diagrams...

Reproduction

Cells can reproduce to make new cells — clever stuff, and it happens in two different ways...

Asexual Reproduction Produces Genetically Identical Cells

1) An ordinary cell can make a new cell by simply dividing in two. The new cell has exactly the same genetic information (i.e. genes) as the parent cell — this is known as asexual reproduction.

In ASEXUAL REPRODUCTION there is only ONE parent, and the offspring has identical genes to the parent (i.e. there's no variation between parent and offspring, so they're clones — see p.29).

2) Here's how it works...

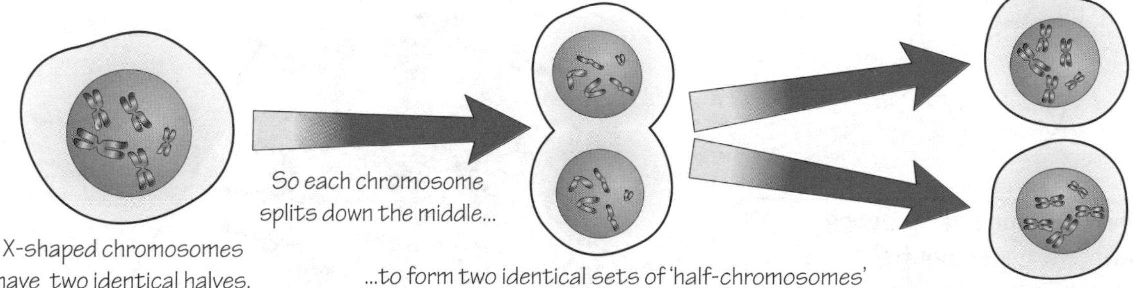

3) This is how all plants and animals grow and produce replacement cells.
4) Some organisms also produce offspring using asexual reproduction, e.g. bacteria and certain plants.

Sexual Reproduction Produces Genetically Different Cells

1) Sexual reproduction is where genetic information from two organisms (a father and a mother) is combined to produce offspring which are genetically different to either parent.
2) In sexual reproduction the mother and father produce gametes — e.g. egg and sperm cells in animals.
3) In humans, each gamete contains 23 chromosomes — half the number of chromosomes in a normal cell. (Instead of having two of each chromosome, a gamete has just one of each.)
4) The egg (from the mother) and the sperm cell (from the father) then fuse together (fertilisation) to form a cell with the full number of chromosomes (half from the father, half from the mother).

SEXUAL REPRODUCTION involves the fusion of male and female gametes. Because there are TWO parents, the offspring contains a mixture of their parents' genes.

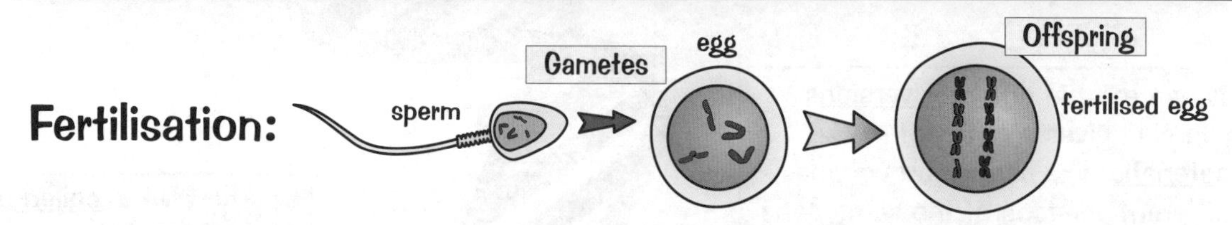

5) This is why the offspring inherits features from both parents — it's received a mixture of chromosomes from its mum and its dad (and it's the chromosomes that decide how you turn out).
6) This is why sexual reproduction produces more variation than asexual reproduction. Pretty cool, eh.

You need to reproduce these facts in the exam...

The main messages on this page are that: 1) asexual reproduction needs just one parent to make genetically identical cells (clones), so there's no variation in the offspring. And 2) sexual reproduction needs two parents to form cells that are genetically different to the parents, so there's lots of variation.

Cloning

We can use asexual reproduction to clone plants and animals in several different ways...

Plants Can Be Cloned from Cuttings and by Tissue Culture

CUTTINGS

1) Gardeners can take cuttings from good parent plants, and then plant them to produce genetically identical copies (clones) of the parent plant.
2) These plants can be produced quickly and cheaply.

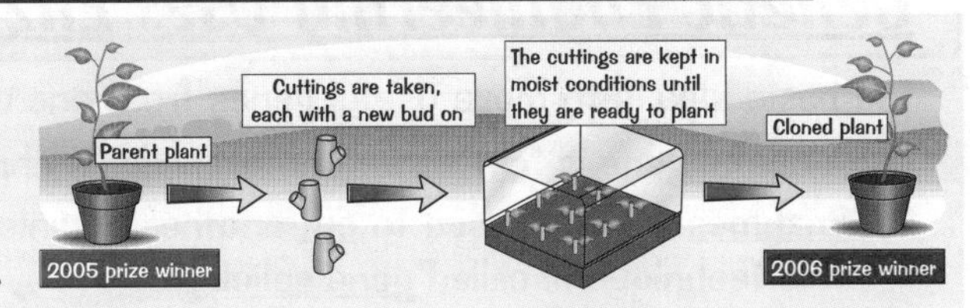

TISSUE CULTURE

This is where a few plant cells are put in a growth medium with hormones, and they then grow into new plants — clones of the parent plant. The advantages of using tissue culture are that you can make new plants very quickly, in very little space, and you can grow all year.

The disadvantage to both these methods is a 'reduced gene pool' (see below).

You Can Make Animal Clones Using Embryo Transplants

Farmers can produce cloned offspring from their best bull and cow — using embryo transplants.

1) Sperm cells are taken from a prize bull and egg cells are taken from a prize cow. The sperm are then used to artificially fertilise an egg cell. The embryo that develops is then split many times (to form clones) before any cells become specialised.
2) These cloned embryos can then be implanted into lots of other cows where they grow into baby calves (which will all be genetically identical to each other).
3) The advantage of this is that hundreds of "ideal" offspring can be produced every year from the best bull and cow.
4) The big disadvantage (as usual) is a reduced gene pool.

A "reduced gene pool" means fewer alleles in a population — which will happen if you breed from the same plants or animals all the time. If a population are all closely related and a new disease appears, all the plants or animals could be wiped out — there may be no allele in the population giving resistance to the disease.

Adult Cell Cloning is Another Way to Make a Clone...

1) Adult cell cloning is the technique that was used to create Dolly — the world-famous cloned sheep.
2) Dolly was made by taking a sheep egg cell and removing its genetic material. A complete set of chromosomes from the cell of an adult sheep was then inserted into the 'empty' egg cell, which then grew into an embryo. This eventually grew into a sheep that was genetically identical to the original adult.
3) Human adult cell cloning could be used to help treat various diseases. A cloned embryo that is genetically identical to the sufferer is created and embryonic stem cells extracted from it. (These can become any cell in the body and could be used to grow replacement cells or organs — without fear of them being rejected by the sufferer's immune system.)
4) Some people think it's unethical to do this as embryos genetically identical to the sufferer are created and then destroyed.
5) Fusion cloning will avoid this problem. Here, an adult cell is fused (joined) to an already existing (but genetically different) embryonic stem cell. The result has the properties of a stem cell but the same genes as the adult.

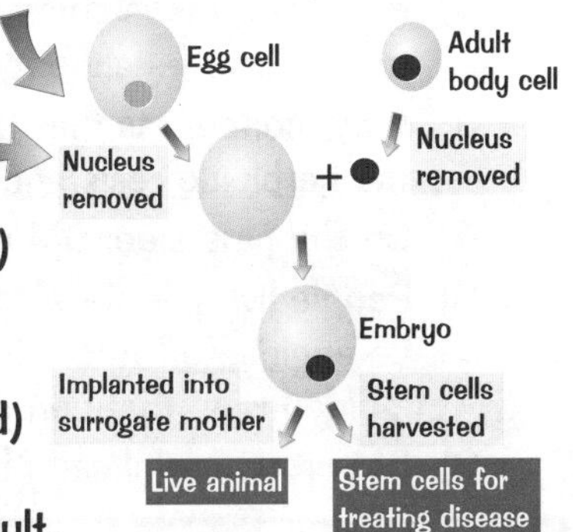

Episode Two — Attack of Dolly the Sheep...

Cloning can be a controversial topic — especially when it's to do with cloning animals (and especially humans). Is it healthy scientific progress, or are we trying to 'play God'?

Genetic Engineering

Scientists can now add, remove or change an organism's genes to alter its characteristics. This is a new science with exciting possibilities, but there might be dangers too...

Genetic Engineering Uses Enzymes to Cut and Paste Genes

The basic idea is to move useful genes from one organism's chromosomes into the cells of another...

1) A useful gene is "cut" from one organism's chromosome using enzymes.
2) Enzymes are then used to cut another organism's chromosome and then to insert the useful gene. This technique is called gene splicing.
3) Scientists use this method to do all sorts of things — for example, the human insulin gene can be inserted into bacteria to produce human insulin:

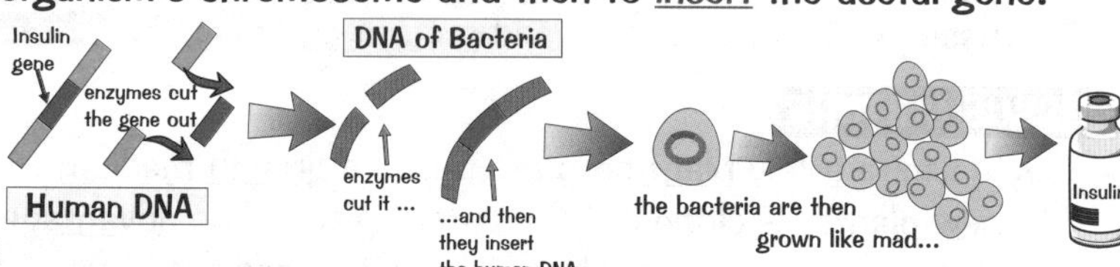

Genes can be Transferred into Animals and Plants

The same method can be used to transfer useful genes into animals and plants at the very early stages of their development (i.e. shortly after fertilisation). This has (or could have) some really useful applications.

1) Genetically modified (GM) plants have been developed that are resistant to viruses and herbicides (chemicals used to kill weeds). And long-life tomatoes can be made by changing the gene that causes the fruit to ripen.
2) Genes can also be inserted into animal embryos so that the animal grows up to have more useful characteristics. For example, sheep have been genetically engineered to produce substances, like drugs, in their milk that can be used to treat human diseases.
3) Genetic disorders like cystic fibrosis are caused by faulty genes. Scientists are trying to cure these disorders by inserting working genes into sufferers. This is called gene therapy.

But Genetic Engineering is a Controversial Topic...

So, genetic engineering is an exciting new area in science which has the potential for solving many of our problems (e.g. treating diseases, more efficient food production etc.) but not everyone thinks it's a great idea.

1) Some people strongly believe that we shouldn't go tinkering about with genes because it's not natural.
2) There are also worries that changing a person's genes might accidentally create unplanned problems — which could then get passed on to future generations.

It's the Same with GM Crops — There Are Pros and Cons...

1) Some people say that growing GM crops will affect the number of weeds and flowers (and therefore wildlife) that usually lives in and around the crops — reducing farmland biodiversity.
2) Not everyone is convinced that GM crops are safe. People are worried they may develop allergies to the food — although there's probably no more risk for this than for eating usual foods.
3) A big concern is that transplanted genes may get out into the natural environment. For example, the herbicide resistance gene may be picked up by weeds, creating a new 'superweed' variety.
4) On the plus side, GM crops can increase the yield of a crop, making more food.
5) People living in developing nations often lack nutrients in their diets. GM crops could be engineered to contain the nutrient that's missing. For example, they're testing 'golden rice' that contains beta-carotene — lack of this substance can cause blindness.
6) GM crops are already being used elsewhere in the world (not the UK) often without any problems.

If only there was a gene to make revision easier...

At the end of the day it's up to the Government to weigh up all the evidence for the pros and cons before making a decision on how this scientific knowledge is used. All scientists can do is make sure the Government has all the information it needs to make the decision.

Evolution

Humans have identified about 1.5 million different species of life on Earth and there's loads more. The highest estimates go up to about 100 million. How did they get here...

No One Knows How Life Began

We know that living things come from other living things — that's easy enough.
But where did the first living thing come from... that's a much more difficult question.

1) There are various theories suggesting how life first came into being, but no one really knows.
2) Maybe the first life forms came into existence in a primordial swamp (or under the sea) here on Earth. Maybe simple organic molecules were brought to Earth on comets — these could have then become more complex organic molecules, and eventually very simple life forms.
3) These suggestions (and others) have been put forward. But we don't know — the evidence has long since been destroyed. All we know is that life started somehow. And from that point on, we're on slightly firmer ground...

The Fossil Record Shows That Organisms Have Evolved

1) A fossil is any evidence of an animal or plant that lived ages ago.
2) Fossils form in rocks as minerals replace slowly decaying tissue (or where no decay happens) and show features like shells, skeletons, soft tissue (occasionally), footprints, etc. They show what was on Earth millions of years ago. They can also give clues about an organism's habitat and the food it ate.
3) We also know that the layers of rock where fossils are found were made at different times. This means it's possible to tell how long ago a particular species lived.
4) From studying the similarities and differences between fossils in differently aged rocks, we can see how species have evolved (changed and developed) over billions of years.

THEORY OF EVOLUTION: Life on Earth began as simple organisms from which all the more complex organisms evolved (rather than just popping into existence).

5) Unfortunately, very few organisms turn into fossils when they die — most just decay away completely. This creates gaps in the fossil record, which means there are many species that no one will ever know about.
6) In theory, you could put all species on a 'family tree' — where each new branch shows the evolution of a new species. Then you could easily find the most recent common ancestor of any two species. The more recent the common ancestor, the more closely related the two species.

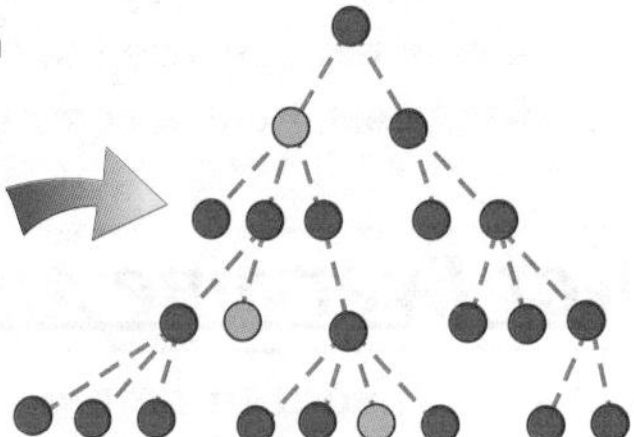

Extinction Happens if You Can't Evolve Quickly Enough

The fossil record contains many species that don't exist any more — these species are said to be extinct. Dinosaurs and mammoths are extinct animals, with only fossils to tell us they existed at all.

Species become extinct for THREE MAIN REASONS:
1) The environment changes too quickly (e.g. destruction of habitat).
2) A new predator or disease kills them all (e.g. humans hunting them).
3) They can't compete with another (new) species for food.

Dodos are now extinct. Humans not only hunted them, but introduced other animals which ate all their eggs, and we destroyed the forest where they lived — they really didn't stand a chance...

Cell, blob, toad, monkey, me — what a fine family tree...

Remember, the most important thing about evolution is that it's a really, really slow process that is happening all the time — there'll always be new species evolving. Also, that 'family tree' is an important idea (even though there'll be bits missing). You can use it to get a picture of how closely related different species are. So humans and chimps would be quite close on the diagram (and have a fairly recent common ancestor), whereas humans and daffodils would be a little further apart.

Evolution

One species evolving into a new species might sound a bit odd, but there's strong evidence that it happens. Scientists also think they have a pretty good understanding of why this happens, and how.

Mutations Are When DNA Changes

(See page 27 for more on genes.)

1) An organism's DNA can change (called a mutation) through everyday wear and tear, e.g. from coming into contact with nasty chemicals. Most of the time mutations have no effect.
2) But, if the mutation happens within a gene AND it's passed on to the next generation it can cause new characteristics. Very occasionally it can give the organism a better chance of survival (e.g. warmer fur, longer legs, larger leaves etc.).
3) Over a long period of time, these 'useful' mutations can help a species adapt to an environment — and eventually may lead to the evolution of a completely new species.

Natural Selection Explains How Evolution Can Occur

1) Charles Darwin came up with the idea of natural selection.
2) He noticed that species tend to be well-adapted to the environment they live in.
3) He argued that organisms that are better adapted have a better chance of survival, and are therefore more likely to breed successfully — passing on their characteristics to the next generation.
4) So if an organism is born with a useful characteristic (either due to the normal shuffling up of mum and dad's genes, or due to mutations), that characteristic has a good chance of being passed on. Whereas if a mutation leads to a disadvantage, the organism may well die before it can breed.
5) So 'good' characteristics accumulate, 'bad' ones are lost. And this is how animals adapt. For example:

Once upon a time maybe all rabbits had short ears and managed OK. Then one day out popped a mutant with BIG EARS who was always the first to hear predators coming and dive for cover. Pretty soon he's had a whole family with BIG EARS, all diving for cover before the other rabbits, and before you know it there's only BIG-EARED rabbits left because the short-eared rabbits just didn't hear trouble coming quick enough.

Not Everyone Agreed with Darwin...

Darwin's theory of natural selection was very controversial at the time — for various reasons...

1. The theory went against common religious beliefs about how life on Earth developed. But Darwin was gradually able to persuade people with his scientific evidence.

2. There were different scientific theories of evolution around at the same time. For example, Lamarck (1744-1829) argued that if a characteristic was used a lot by an organism then it would become more developed — and that these developed characteristics would be passed on to the next generation. So if a rabbit looked a lot it would develop good eyes, and this would mean its offspring would have good eyes too. (Darwin's theory was different because he argued that a rabbit might by chance be born with good eyes that would help it spot predators.)

3. Darwin couldn't give a good explanation about why these new, useful characteristics appeared (but then he didn't know anything about genes or mutations).

You'll increase your chances of survival by learning this stuff...

Darwin was ridiculed by the Church about his theory, but it wasn't the first time a scientist or philosopher was picked on by the Church... Galileo was put under house arrest by the Church for nine years for supporting Copernicus' theory that the Earth was not the centre of the Universe.

Revision Summary 1 for Biology 1b

There's a lot to remember from this part of the section — the stuff about genes, chromosomes and DNA is particularly tricky, so make sure you get your head around it. Cloning and genetic engineering are controversial topics that there's a lot of debate about — and you need to know all sides of the stories. Anyway, the best way to check you've got it all is to try these questions... and keep trying them until you've got them sussed.

1)* Big penguins like Phyllis can quite happily live in the coldest parts of the Antarctic. Smaller types of penguin like Arnold live in warmer areas like the south of South America.

Phyllis

Arnold

a) Explain why this is.
b) Suggest some other features that Phyllis is likely to have to allow her to survive in the Antarctic.

2)* The kangaroo rat lives in the hot deserts of North America. What adaptations would you expect the kangaroo rat to have? Think about the following:
- urine production and concentration.
- sweating
- surface area

3) How are desert plants adapted to their environment?
4) State three ways that plants and animals might be adapted to deter predators.
5) What are the three main factors that determine the size of a population?
6) Name three things that: a) plants compete for, b) animals compete for.
7)* This graph shows the population size of both red and grey squirrels for the UK.

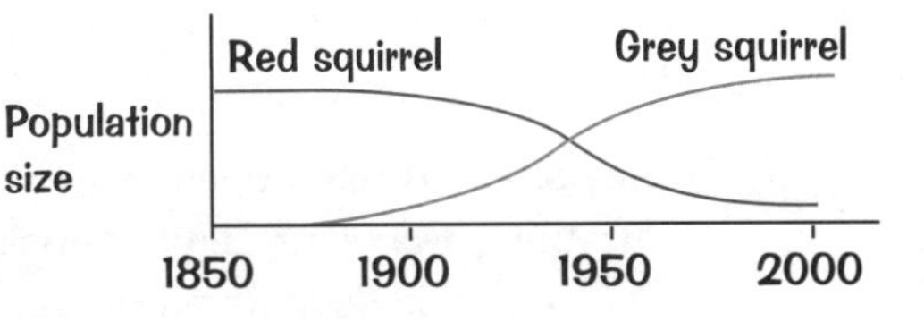

a) In what year were grey squirrels introduced to the UK?
b) What happened to the population of red squirrels after grey squirrels were introduced? Suggest why this happened.
c) Name two things that the two species may compete for.

8) What are the two types of variation?
9) List four features of animals which aren't affected at all by the environment, and four which are.
10) Draw a set of diagrams showing the relationship between: cell, nucleus, chromosomes, DNA.
11) How many pairs of chromosomes does a normal human cell nucleus contain? What cells, found in every adult human, are the exception to this?
12) Give the definition of asexual reproduction.
13) Explain why sexual reproduction results in offspring that are genetically different from either parent.
14) Describe how to make plant clones from: a) cuttings, b) tissue culture.
15) Give an advantage and a disadvantage of producing cloned plants.
16) Describe two ways to clone an animal.
17) What is used to cut genes from a chromosome in genetic engineering?
18) Give an example of how genetic engineering has been used to help people.
19) Give a balanced account of some of the views that different people have about genetic engineering.
20) Give three reasons why species become extinct.
21) Suggest how a deer-like animal might evolve to have longer legs than it once had.
22) a) Describe Darwin's theory of evolution. Why was this theory once very controversial?
b)* If this is just a theory that can't be proved for certain, why do so many people accept it?

* Answers on page 96

Human Impact on the Environment

We have an impact on the world around us — and the more humans there are, the bigger the impact.

There are Six Billion People in the World...

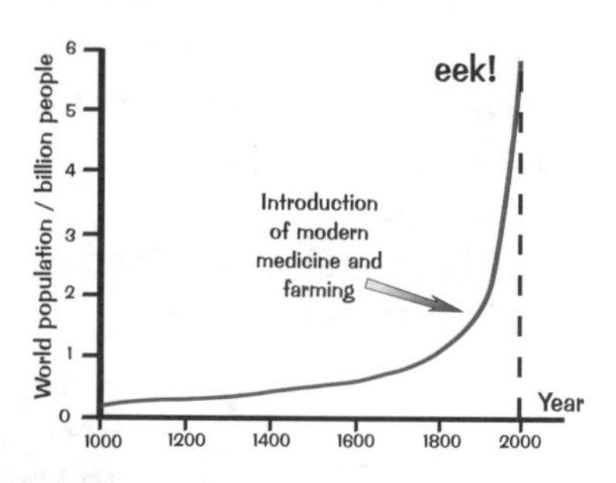

1) The population of the world is currently rising very quickly, and it's not slowing down — look at the graph...
2) This is mostly due to modern medicine and farming methods, which have reduced the number of people dying from disease and hunger.
3) This is great for all of us humans, but it means we're having a bigger effect on the environment we live in...

...With Increasing Demands on the Environment

When the Earth's population was much smaller, the effects of human activity were usually small and local. Nowadays though, our actions can have a far more widespread effect.

1) Our rapidly increasing population puts pressure on the environment, as we take the resources we need to survive.
2) But people around the world are also demanding a higher standard of living (and so demand luxuries to make life more comfortable — cars, computers, etc.). So we use more raw materials (e.g. oil to make plastics), but we also use more energy for the manufacturing processes. This all means we're taking more and more resources from the environment more and more quickly.
3) Unfortunately, many raw materials are being used up quicker than they're being replaced. So if we carry on like we are, one day we're going to run out.

We're Also Producing More Waste

As we make more and more things we produce more and more waste. And unless this waste is properly handled, more harmful pollution will be caused. This affects water, land and air.

Water Sewage and toxic chemicals from industry can pollute lakes, rivers and oceans, affecting the plants and animals that rely on them for survival (including humans). And the chemicals used on land (e.g. fertilisers) can be washed into water.

Land We use toxic chemicals for farming (e.g. pesticides and herbicides). We also bury nuclear waste underground, and we dump a lot of household waste in landfill sites.

Air Smoke and gases released into the atmosphere can pollute the air (see page 35 for more info). For example, sulfur dioxide can cause acid rain.

More People Means Less Land for Plants and Other Animals

Humans also reduce the amount of land and resources available to other animals and plants. The four main human activities that do this are:

1) Building

2) Farming

3) Dumping Waste

4) Quarrying

More people, more mess, less space, less resources...

Well, I feel guilty, I don't know about you. First we kill off loads of animals by hunting them, then we try to manipulate their DNA, and now we're destroying the land they live on. In the exam you might be given some data about environmental impact, so make sure you understand what's going on...

The Greenhouse Effect

The greenhouse effect is always in the news. We need it, since it makes Earth a suitable temperature for living on. But it's starting to trap more heat than is necessary.

Carbon Dioxide and Methane Trap Heat from the Sun

1) The temperature of the Earth is a balance between the heat it gets from the Sun and the heat it radiates back out into space.
2) Gases in the atmosphere naturally act like an insulating layer. They absorb most of the heat that would normally be radiated out into space, and re-radiate it in all directions (including back towards the Earth).

This is what happens in a greenhouse. The sun shines in, and the glass helps keeps some of the heat in.

3) If this didn't happen, then at night there'd be nothing to keep any heat in, and we'd quickly get very cold indeed. But recently we've started to worry that this effect is getting a bit out of hand...

4) There are several different gases in the atmosphere which help keep the heat in. They're called "greenhouse gases" (oddly enough) and the main ones whose levels we worry about are carbon dioxide and methane — because the levels of these two gases are rising quite sharply.
5) The Earth is gradually heating up because of the increasing levels of greenhouse gases — this is global warming. Global warming is a type of climate change and causes other types of climate change, e.g. changing rainfall patterns.

Human Activity Produces Lots of Carbon Dioxide

1) Humans release carbon dioxide into the atmosphere all the time as part of our everyday lives — in car exhausts, industrial processes, as we burn fossil fuels etc.
2) People around the world are also cutting down large areas of forest (deforestation) for timber and to clear land for farming — and this activity affects the level of carbon dioxide in the atmosphere in various ways:

- Carbon dioxide is released when trees are burnt to clear land. (Carbon in wood is 'locked up' and doesn't contribute to atmospheric pollution — until it's released by burning.)
- Microorganisms feeding on bits of dead wood release CO_2 as a waste product of respiration.
- Cutting down loads of trees means that the amount of carbon dioxide removed from the atmosphere during photosynthesis is reduced.

So we're putting more CO_2 into the atmosphere and taking less out.

Methane is Also a Problem...

1) Methane gas is also contributing to the greenhouse effect.
2) It's produced naturally from various sources, e.g. rotting plants in marshland.
3) However, two 'man-made' sources of methane are on the increase:
   a) Rice growing
   b) Cattle rearing — it's the cows' "pumping" that's the problem, believe it or not.

Methane is a stinky problem but an important one...

Global warming is rarely out of the news. Scientists accept that it's happening and that human activity has caused most of the recent warming. However, they don't know exactly what the effects will be...

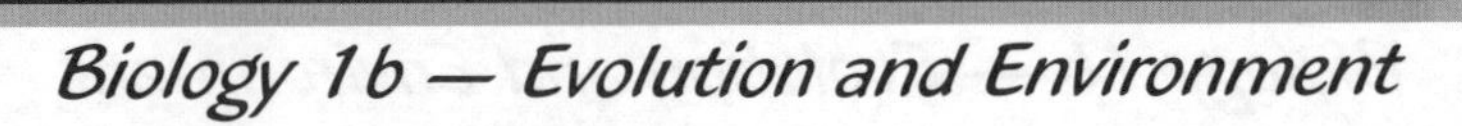

Climate Change

The Earth is getting warmer. Climate scientists are now trying to work out what the effects of global warming might be — sadly, it's not as simple as everyone having nicer summers.

The Consequences of Global Warming Could be Pretty Serious

There are several reasons to be worried about global warming. Here are a few:

1) As the sea gets warmer, it expands, causing sea level to rise. Sea level has risen a little bit over the last 100 years. If it keeps rising it'll be bad news for people living in low-lying places like the Netherlands, East Anglia and the Maldives — they'd be flooded.
2) Higher temperatures make ice melt. Water that's currently 'trapped' on land (as ice) runs into the sea, causing sea level to rise even more.
3) Global warming has changed weather patterns in many parts of the world. It's thought that many regions will suffer more extreme weather because of this, e.g. longer, hotter droughts. Hurricanes form over water that's warmer than 27 °C — so with more warm water, you'd expect more hurricanes.
4) Changing weather patterns also affect food production — some regions are now too dry to grow food, some too wet. This will get worse as temperature increases and weather patterns change more.
5) The climate is a very complicated system. For instance, if the ice melts, there's less white stuff around to reflect the sun's rays out to space, so maybe we'll absorb more heat and get even warmer. But... when the sea's warmer, more water evaporates, making more clouds — and they reflect the Sun's rays, so maybe we'd cool down again. So it's hard to predict exactly what will happen, but lots of people are working on it, and it's not looking too good.

You Need to Weigh the Evidence Before Making Judgements

1) To find out how our climate is changing, scientists are busy collecting data about the environment.
2) For instance, we're using satellites to monitor snow and ice cover, and to measure the temperature of the sea surface. We're recording the temperature and speed of the ocean currents, to try and detect any changes. Automatic weather stations are constantly recording atmospheric temperatures.
3) All this data is only useful if it covers a wide enough area and a long enough time scale.
4) Generally, observations of a very small area aren't much use. Noticing that your local glacier seems to be melting does not mean that ice everywhere is melting, and it's certainly not a valid way to show that global temperature is changing. (That would be like going to Wales, seeing a stripy cow and concluding that all the cows in Wales are turning into zebras.) Looking at the area of ice cover over a whole continent, like Antarctica, would be better.

5) The same thing goes for time. It's no good going to the Arctic, seeing four polar bears one week but only two the next week and concluding that polar bears are dying out because the ice is disappearing. You need to do your observations again and again, year after year.
6) Scientists can make mistakes — so don't take one person's word for something, even if they've got a PhD. But if lots of scientists get the same result using different methods, it's probably right. That's why most governments around the world are starting to take climate change seriously.

Climate control — it's optional on most 4×4s...

We humans have created some big environmental problems for ourselves. Many people, and some governments, think we ought to start cleaning up the mess. Scientists can help, mainly in understanding the problems and suggesting solutions, but it's society as a whole that has to do something.

Sustainable Development

There is a growing feeling among scientists and politicians that if we carry on behaving as we are, we may end up causing huge problems for future generations...

Sustainable Development Needs Careful Planning

1) Human activities can damage the environment (e.g. pollution). And some of the damage we do can't easily be repaired (e.g. the destruction of the rainforests).
2) We're also placing greater pressure on our planet's limited resources (e.g. oil is a non-renewable resource so it will eventually run out).
3) This means that we need to plan carefully to make sure that our activities today don't mess things up for future generations — this is the idea behind sustainable development...

SUSTAINABLE DEVELOPMENT meets the needs of today's population without harming the ability of future generations to meet their own needs.

4) This isn't easy — it needs detailed thought at every level to make it happen. For example, governments around the world will need to make careful plans. But so will the people in charge at a regional level.

Reduction in Biodiversity Could Be a Big Problem

Biodiversity is the variety of different species present in an area — the more species, the higher the biodiversity. Ecosystems (especially tropical rainforests) can contain a huge number of different species, so when a habitat is destroyed there is a danger of many species becoming extinct — biodiversity is reduced. This causes a number of lost opportunities for humans and problems for those species that are left:

1) There are probably loads of useful products that we will never know about because the organisms that produced them have become extinct. Newly discovered plants and animals are a great source of new foods, new fibres for clothing and new medicines, e.g. the rosy periwinkle flower from Madagascar has helped treat Hodgkin's disease (a type of cancer), and a chemical in the saliva of a leech has been used to help prevent blood clots during surgery.
2) Loss of one or more species from an ecosystem unbalances it, e.g. the extinct animal's predators may die out or be reduced. Loss of biodiversity can have a 'snowball effect' which prevents the ecosystem providing things we need, such as rich soil, clean water, and the oxygen we breathe.

Human Impact can be Measured Using Indicator Species

Getting an accurate picture of the human impact on the environment is hard. But one technique that's used involves indicator species.

1) Some organisms are very sensitive to changes in their environment and so can be studied to see the effect of human activities — these organisms are known as indicator species.

2) For example, air pollution can be monitored by looking at particular types of lichen, which are very sensitive to levels of sulfur dioxide in the atmosphere (and so can give a good idea about the level of pollution from car exhausts, power stations, etc.). The number and type of lichen at a particular location will indicate how clean the air is (e.g. the air is clean if there are lots of lichen).
3) And if raw sewage is released into a river, the bacterial population in the water increases and uses up the oxygen. Animals like mayfly larvae are good indicators for water pollution, because they are very sensitive to the level of oxygen in the water. If you find mayfly larvae in a river, it indicates that the water is clean.

Teenagers are an indicator species — not found in clean rooms...

In the exam, make sure you remember the details about the environmental problems that development can cause. If you get an essay-type question, stick 'em in to show off your 'scientific knowledge'.

Revision Summary 2 for Biology 1b

Eeee... there's no cheerful stuff here. It's all doom and gloom. All these humans are hogging the land, and filling the air with bad stuff. In fact, planet Earth would be much better off if we all emigrated to Mars.

Anyway, it's time for you to answer some questions on the woes of the world. You should know the drill by now. Try all the questions and if you get any wrong it's straight back to the page on that topic — do not pass go, do not collect £200. Before long, you'll be a guru on the environment and won't be fazed by any question that turns up in the exam.

1) What is happening to the world's population? What is largely responsible for this trend?
2) Suggest three ways in which a rising population is affecting the environment.
3) What are the main four human activities that use up land?
4) Name two important greenhouse gases. Why are they called 'greenhouse' gases?
5) Draw and label a diagram to explain the greenhouse effect.
6) Give three ways that deforestation adds to the greenhouse effect.
7) Give two human activities (apart from deforestation) that release carbon dioxide.
8) Name two things that are increasing and releasing more methane.
9) What problems could global warming cause?
10)* Read the statement below and consider how valid it is.

    The Malaspina Glacier in Alaska is losing over 2.7 km³ of water each year.
    This proves that global warming is happening.

11) Define sustainable development.
12)* The graph on the right shows population growth and an estimate of the number of species that have become extinct between 1800 and 2000.
    a) How are human population and the number of extinct species related?
    b) Suggest a reason for this relationship.

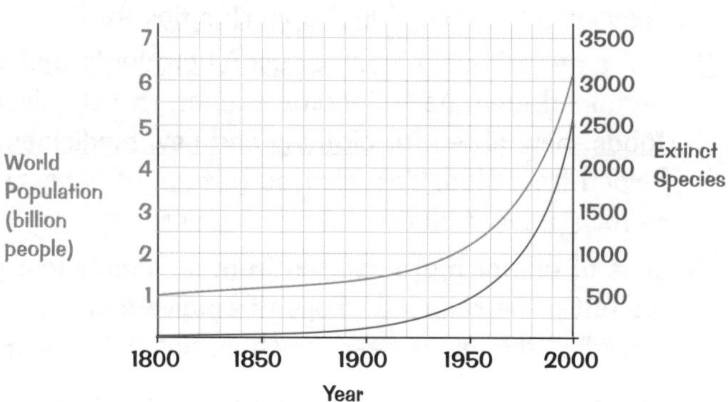

13) Explain how lichen can be used as an indicator of air pollution.
14) What does it mean if you find mayfly larvae in a river?
    a) it's May, b) the water is clean, c) the water is dirty
15)* Here is a graph of population and household waste (destined for landfill) produced for a small village.
    a) In 2002 how many people lived in the village?
    b) In what year did the village produce 12.5 tonnes of waste?
    c) Using your answer from part a), work out how many tonnes one person produced in 2002.
    d) What year did the village bring in a recycling scheme?
    e) Give two ways that a recycling scheme could benefit the environment.

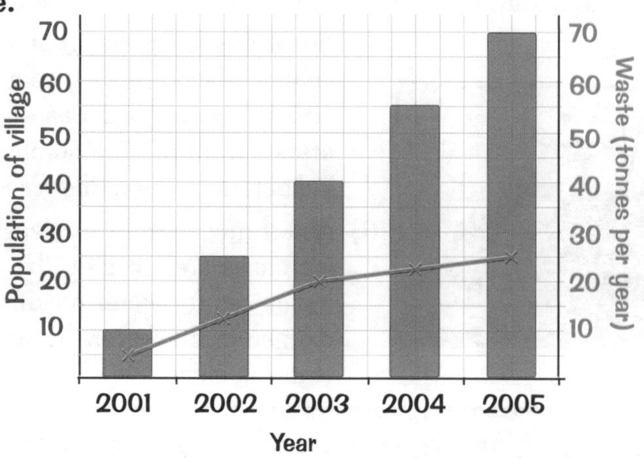

* Answers on page 96

Life and Cells

In physics you often start off with forces, in chemistry it's usually elements, and in biology it's the cell. Not very original, but nice and familiar at least. So away we go — a-one, a-two, a-one, two, three, four...

Plant and Animal Cells have Similarities and Differences

Most human cells, like most animal cells, have the following parts — make sure you know them all:

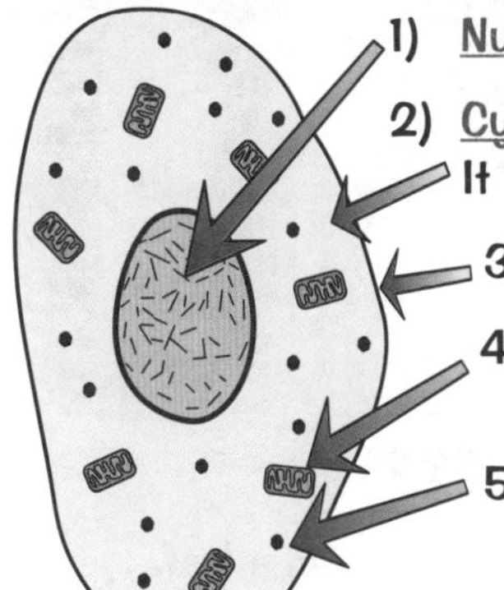

1) Nucleus — contains genetic material that controls the activities of the cell.
2) Cytoplasm — gel-like substance where most of the chemical reactions happen. It contains enzymes (see page 53) that control these chemical reactions.
3) Cell membrane — holds the cell together and controls what goes in and out.
4) Mitochondria — these are where most of the reactions for respiration take place (see page 54). Respiration releases energy that the cell needs to work.
5) Ribosomes — these are where proteins are made in the cell.

Plant cells usually have all the bits that animal cells have, plus a few extra things that animal cells don't have:

1) Rigid cell wall — made of cellulose. It supports the cell and strengthens it.
2) Permanent vacuole — contains cell sap, a weak solution of sugar and salts.
3) Chloroplasts — these are where photosynthesis occurs, which makes food for the plant (see page 43). They contain a green substance called chlorophyll.

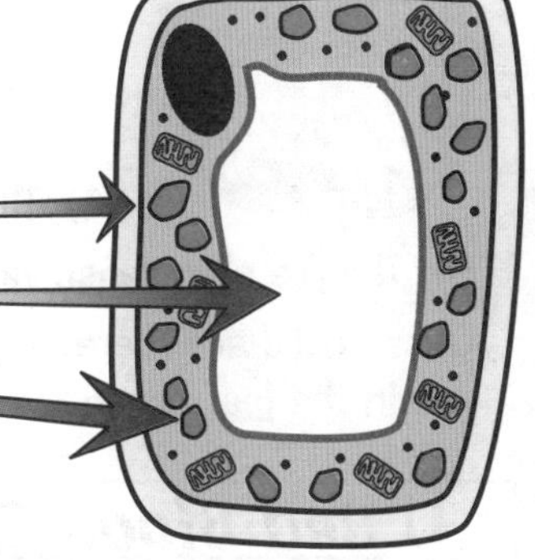

Cells Make Up Tissues, Organs and Systems

Cells have structures that are specialised so they can carry out their function (see next page). Similar cells are grouped together to make a tissue, and different tissues work together as an organ. Organs have a particular job to do in the body — e.g. the heart circulates the blood. Groups of organs working together make up an organ system, like the digestive system. And finally, groups of organs and organ systems working together make up a full organism like you or me. Phew, it's pretty complicated, this life business.

Here's a plant example, but there are loads of animal examples as well of course:

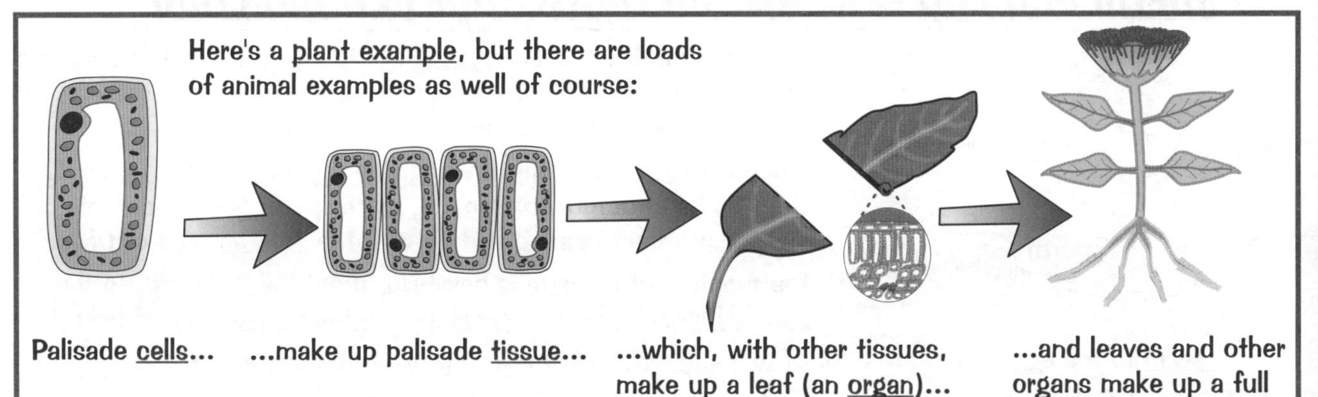

Palisade cells... ...make up palisade tissue... ...which, with other tissues, make up a leaf (an organ)... ...and leaves and other organs make up a full plant (an organism).

There's quite a bit to learn in biology — but that's life, I guess...

At the top of the page are typical cells with all the typical bits you need to know. But cells aren't all the same — they have different structures and produce different substances depending on the job they do.

Specialised Cells

Most cells are specialised for their specific function within a tissue or organ. In the exam you might have to explain how a particular cell is adapted for its function. Here are a few examples that might come up:

1) Palisade Leaf Cells Are Adapted for Photosynthesis

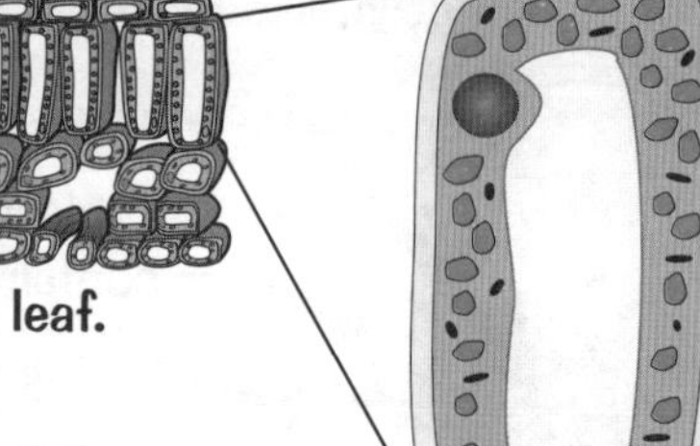

1) Packed with chloroplasts for photosynthesis. More of them are crammed at the top of the cell — so they're nearer the light.
2) Tall shape means a lot of surface area exposed down the side for absorbing CO_2 from the air in the leaf.
3) Thin shape means that you can pack loads of them in at the top of a leaf.

Palisade leaf cells are grouped together to give the palisade layer of a leaf — this is the leaf tissue where most of the photosynthesis happens.

2) Guard Cells Are Adapted to Open and Close Pores

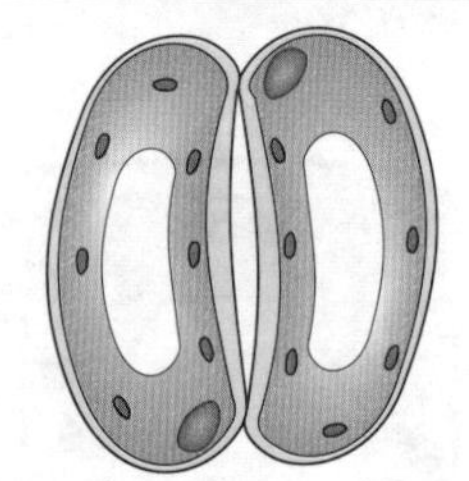

1) Special kidney shape which opens and closes the stomata (pores) in a leaf.
2) When the plant has lots of water the guard cells fill with it and go plump and turgid. This makes the stomata open so gases can be exchanged for photosynthesis.
3) When the plant is short of water, the guard cells lose water and become flaccid, making the stomata close. This helps stop too much water vapour escaping.
4) Thin outer walls and thickened inner walls make the opening and closing work.
5) They're also sensitive to light and close at night to save water without losing out on photosynthesis.

Guard cells are therefore adapted to their function of allowing gas exchange and controlling water loss within the leaf organ.

3) Red Blood Cells Are Adapted to Carry Oxygen

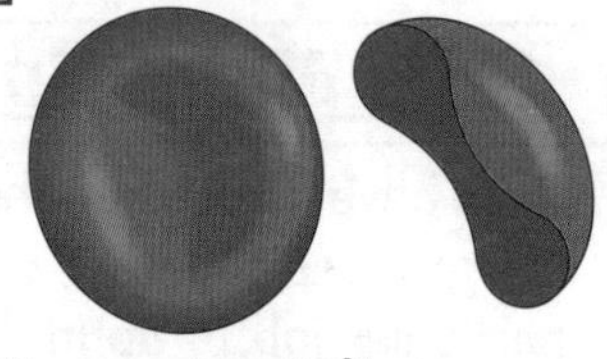

1) Concave shape gives a big surface area for absorbing oxygen. It also helps them pass smoothly through capillaries to reach body cells.
2) They're packed with haemoglobin — the pigment that absorbs the oxygen.
3) They have no nucleus, to leave even more room for haemoglobin.

Red blood cells are an important part of the blood (blood's actually counted as a tissue — weird).

4) Sperm and Egg Cells Are Specialised for Reproduction

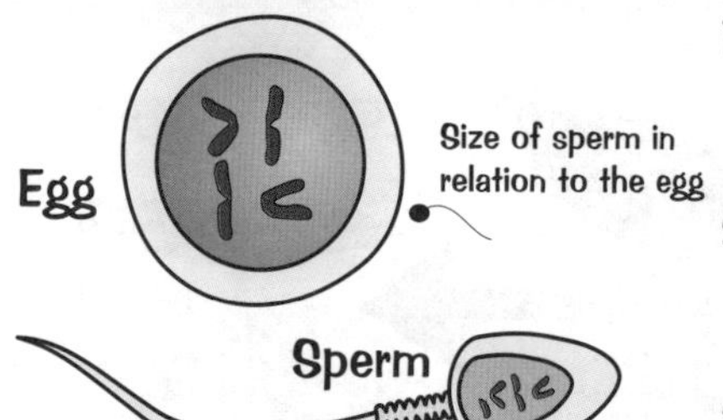

1) The main functions of an egg cell are to carry the female DNA and to nourish the developing embryo in the early stages. The egg cell contains huge food reserves to feed the embryo.
2) When a sperm fuses with the egg, the egg's membrane instantly changes its structure to stop any more sperm getting in. This makes sure the offspring end up with the right amount of DNA.
3) The function of a sperm is basically to get the male DNA to the female DNA. It has a long tail and a streamlined head to help it swim to the egg. There are a lot of mitochondria in the cell to provide the energy needed.
4) Sperm also carry enzymes in their heads to digest through the egg cell membrane.

Sperm and eggs are very important cells in the reproductive system.

Beans, flying saucers, tadpoles — cells are masters of disguise...

Okay so the red blood cell doesn't have a nucleus, but apart from that these cells all have all the bits you learnt about on page 39, even though they look completely different and do totally different jobs.

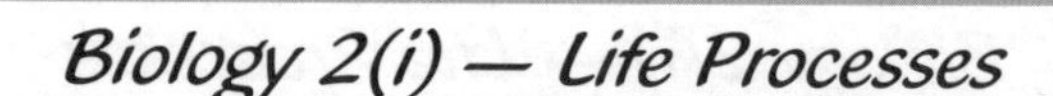

Diffusion

Particles move about randomly, and after a bit they end up evenly spaced. It's not rocket science, is it...

Don't Be Put Off by the Fancy Word

"Diffusion" is simple. It's just the gradual movement of particles from places where there are lots of them to places where there are fewer of them. That's all it is — just the natural tendency for stuff to spread out. Unfortunately you also have to learn the fancy way of saying the same thing, which is this:

DIFFUSION is the passive movement of particles from an area of HIGH CONCENTRATION to an area of LOW CONCENTRATION

Diffusion happens in both liquids and gases — that's because the particles in these substances are free to move about randomly. The simplest type is when different gases diffuse through each other. This is what's happening when the smell of perfume diffuses through a room:

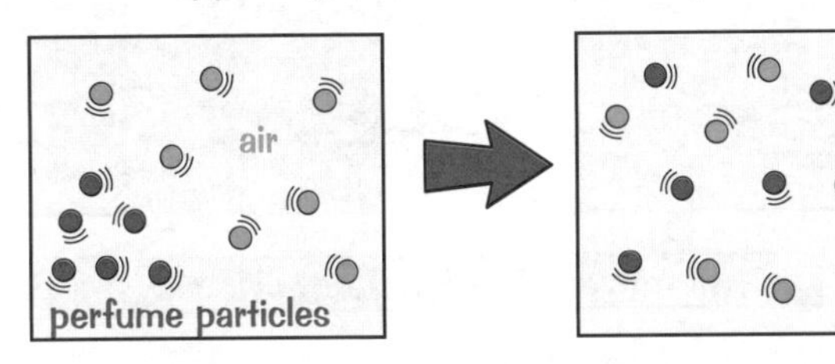

The bigger the difference in concentration, the faster the diffusion rate.

Cell Membranes Are Kind of Clever...

They're clever because they hold the cell together BUT they let stuff in and out as well. Substances can move in and out of cells by diffusion and osmosis (see next page). Only very small molecules can diffuse through cell membranes though — things like glucose, amino acids, water and oxygen.
Big molecules like starch and proteins can't fit through the membrane.

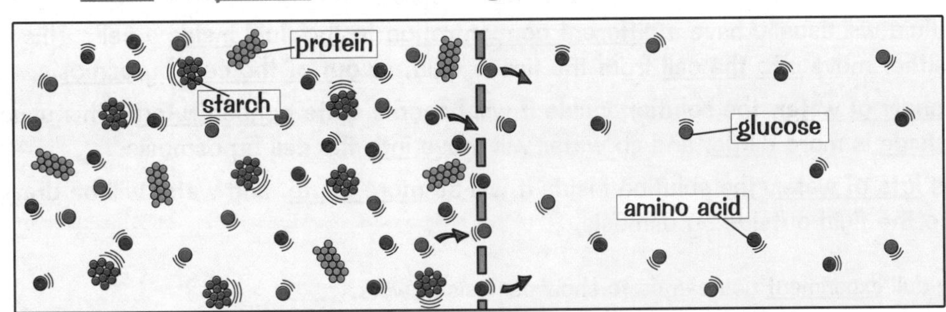

1) Just like with diffusion in air, particles flow through the cell membrane from where there's a high concentration (a lot of them) to where there's a low concentration (not such a lot of them).
2) They're only moving about randomly of course, so they go both ways — but if there are a lot more particles on one side of the membrane, there's a net (overall) movement from that side.
3) The rate of diffusion depends on three main things:
   a) Distance — substances diffuse more quickly when they haven't as far to move. Pretty obvious.
   b) Concentration difference (gradient) — substances diffuse faster if there's a big difference in concentration. If there are lots more particles on one side, there are more there to move across.
   c) Surface area — the more surface there is available for molecules to move across, the faster they can get from one side to the other.

Revision by diffusion — you wish ...

Wouldn't that be great — if all the ideas in this book would just gradually drift across into your mind, from an area of high concentration (in the book) to an area of low concentration (in your mind — no offence). Actually, that probably will happen if you read it again. Why don't you give it a go...

Osmosis

If you've got your head round diffusion, osmosis will be a breeze. If not, what are you doing turning over?

Osmosis is a Special Case of Diffusion, That's All

OSMOSIS is the movement of water molecules across a partially permeable membrane from a region of high water concentration to a region of low water concentration.

1) A partially permeable membrane is just one with very small holes in it. So small, in fact, only tiny molecules (like water) can pass through them, and bigger molecules (e.g. sucrose) can't.
2) The water molecules actually pass both ways through the membrane during osmosis. This happens because water molecules move about randomly all the time.
3) But because there are more water molecules on one side than on the other, there's a steady net flow of water into the region with fewer water molecules, i.e. into the stronger sugar solution.
4) This means the strong sugar solution gets more dilute. The water acts like it's trying to "even up" the concentration either side of the membrane.

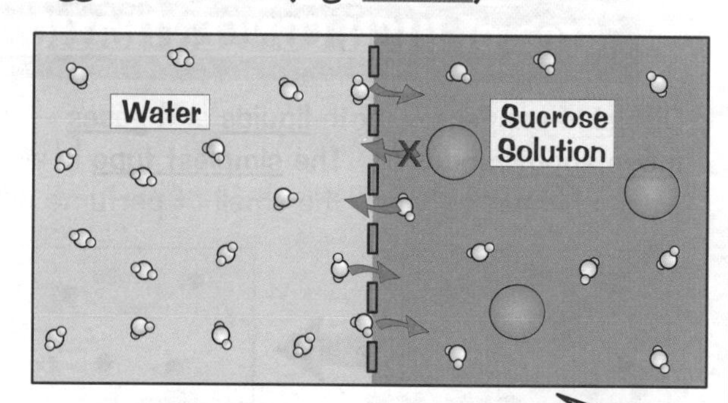

5) Osmosis is a type of diffusion — passive movement of water particles from an area of high water concentration to an area of low water concentration.

Water Moves Into and Out of Cells by Osmosis

1) Tissue fluid surrounds the cells in the body — it's basically just water with oxygen, glucose and stuff dissolved in it. It's squeezed out of the blood capillaries to supply the cells with everything they need.
2) The tissue fluid will usually have a different concentration to the fluid inside a cell. This means that water will either move into the cell from the tissue fluid, or out of the cell, by osmosis.
3) If a cell is short of water, the solution inside it will become quite concentrated. This usually means the solution outside is more dilute, and so water will move into the cell by osmosis.
4) If a cell has lots of water, the solution inside it will be more dilute, and water will be drawn out of the cell and into the fluid outside by osmosis.

There's a fairly dull experiment you can do to show osmosis at work.

You cut up an innocent potato into identical cylinders, and get some beakers with different sugar solutions in them. One should be pure water, another should be a very concentrated sugar solution. Then you can have a few others with concentrations in between.

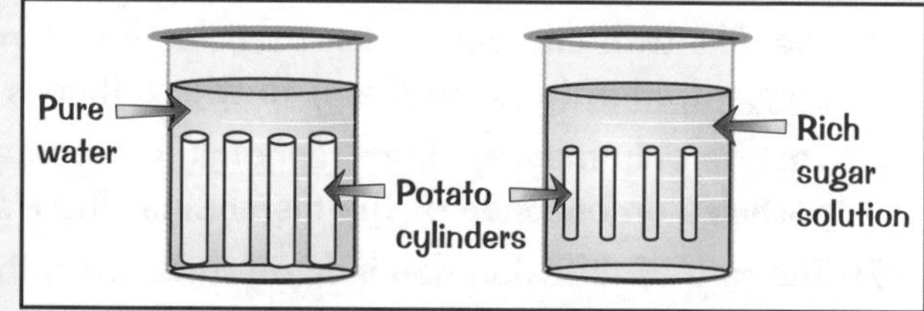

You measure the length of the cylinders, then leave a few cylinders in each beaker for half an hour or so. Then you take them out and measure their lengths again. If the cylinders have drawn in water by osmosis, they'll be a bit longer. If water has been drawn out, they'll have shrunk a bit. Then you can plot a few graphs and things.

The dependent variable is the chip length and the independent variable is the concentration of the sugar solution. All other variables (volume of solution, temperature, time, type of sugar used, etc. etc.) must be kept the same in each case or the experiment won't be a fair test. See, told you it was dull.

And to all you cold-hearted potato murderers...

And that's why it's bad to drink sea-water. The high salt content means you end up with a much lower water concentration in your blood and tissue fluid than in your cells. All the water is sucked out of your cells by osmosis and they shrivel and die. So next time you're stranded at sea, remember this page...

Photosynthesis

You must learn the photosynthesis equation. Learn it so well that you'll still remember it when you're 109.

Learn the Equation for Photosynthesis:

$$\text{Carbon dioxide} + \text{water} \xrightarrow[\text{chlorophyll}]{\text{SUNLIGHT}} \text{glucose} + \text{oxygen}$$

Photosynthesis Produces Glucose Using Sunlight

1) Photosynthesis is the process that produces 'food' in plants. The 'food' it produces is glucose.
2) Photosynthesis happens in the leaves of all green plants — this is largely what the leaves are for.
3) Photosynthesis happens inside the chloroplasts, which are found in leaf cells and in other green parts of a plant. Chloroplasts contain a substance called chlorophyll, which absorbs sunlight and uses its energy to convert carbon dioxide and water into glucose. Oxygen is also produced.

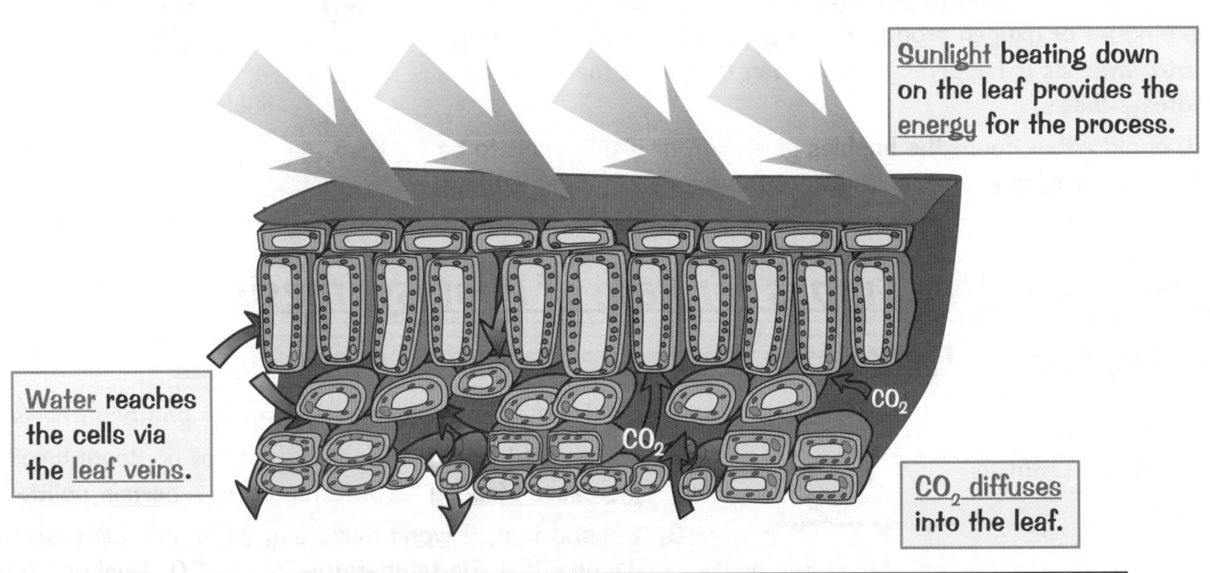

Four Things are Needed for Photosynthesis to Happen:

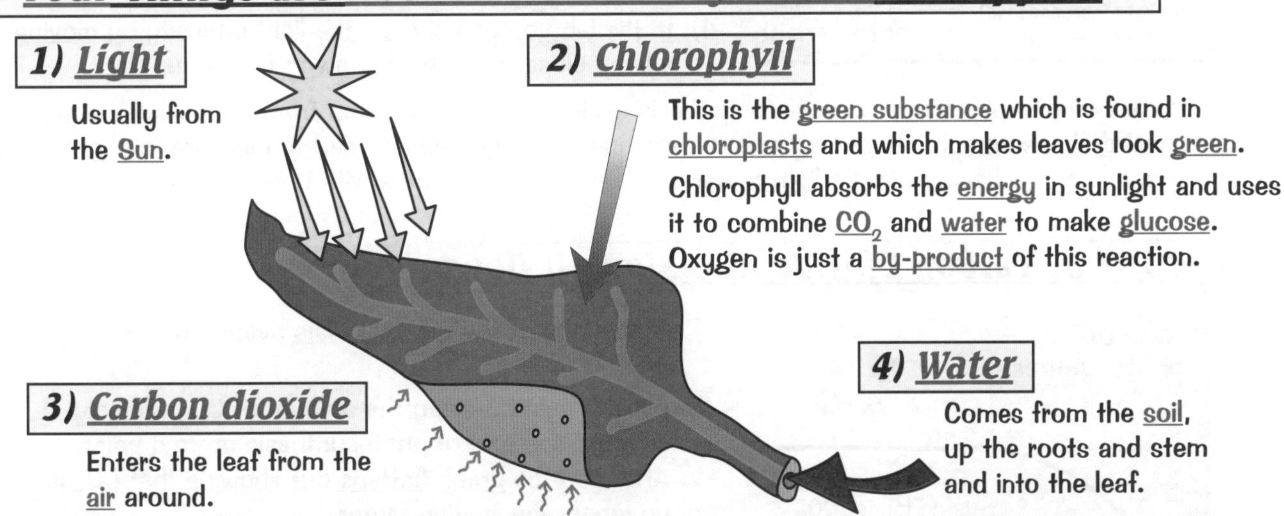

Now you'll have something to bore the great-grandkids with...

You'll be able to tell them how in your day, all you needed was a bit of carbon dioxide and some water and you could make your own entertainment. See, when you're 109 you're allowed to get a bit confused, but in the middle of an exam you most certainly are not. So if you don't know it, learn it. (And if you do, learn it again anyway.)

The Rate of Photosynthesis

The rate of photosynthesis is affected by the amount of light, the amount of CO_2, and the temperature. Plants also need water for photosynthesis, but when a plant is so short of water that it becomes the limiting factor in photosynthesis, it's already in such trouble that this is the least of its worries.

The Limiting Factor Depends on the Conditions

1) Any of these three factors can become the limiting factor. This just means that it's stopping photosynthesis from happening any faster.
2) Which factor is limiting at a particular time depends on the environmental conditions:
   - at night it's pretty obvious that light is the limiting factor,
   - in winter it's often the temperature,
   - if it's warm enough and bright enough, the amount of CO_2 is usually limiting.

You can do experiments to work out the ideal conditions for photosynthesis in a particular plant. The easiest type to use is a water plant like Canadian pondweed — you can easily measure the amount of oxygen produced in a given time to show how fast photosynthesis is happening (remember, oxygen is made during photosynthesis).
You could either count the bubbles given off, or if you want to be a bit more accurate you could collect the oxygen in a gas syringe.

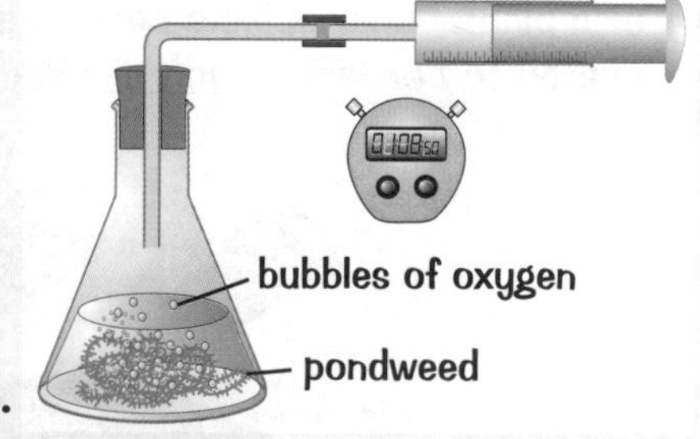

Three Important Graphs for Rate of Photosynthesis

1) Not Enough Light Slows Down the Rate of Photosynthesis

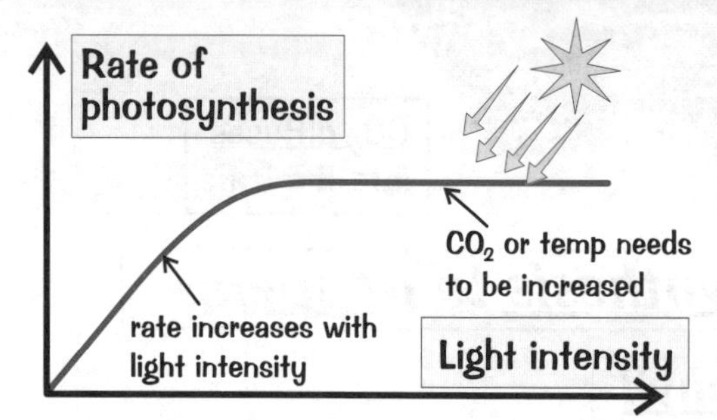

1) Light provides the energy needed for photosynthesis.
2) As the light level is raised, the rate of photosynthesis increases steadily — but only up to a certain point.
3) Beyond that, it won't make any difference because then it'll be either the temperature or the CO_2 level which is the limiting factor.
4) In the lab you can change the light intensity by moving a lamp closer to or further away from your plant.
5) But if you just plot the rate of photosynthesis against "distance of lamp from the beaker", you get a weird-shaped graph. To get a graph like the one above you either need to measure the light intensity at the beaker using a light meter or do a bit of nifty maths with your results.

2) Too Little Carbon Dioxide Also Slows it Down

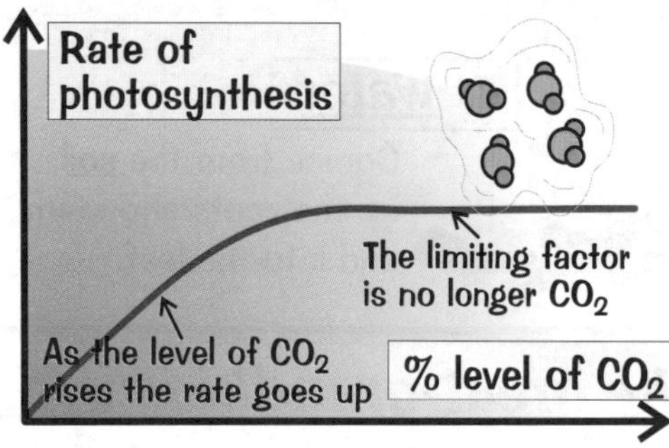

1) CO_2 is one of the raw materials needed for photosynthesis.
2) As with light intensity the amount of CO_2 will only increase the rate of photosynthesis up to a point. After this the graph flattens out showing that CO_2 is no longer the limiting factor.
3) As long as light and CO_2 are in plentiful supply then the factor limiting photosynthesis must be temperature.
4) There are loads of different ways to control the amount of CO_2. One way is to dissolve different amounts of sodium hydrogencarbonate in the water, which gives off CO_2.

The Rate of Photosynthesis

3) The Temperature has to be Just Right

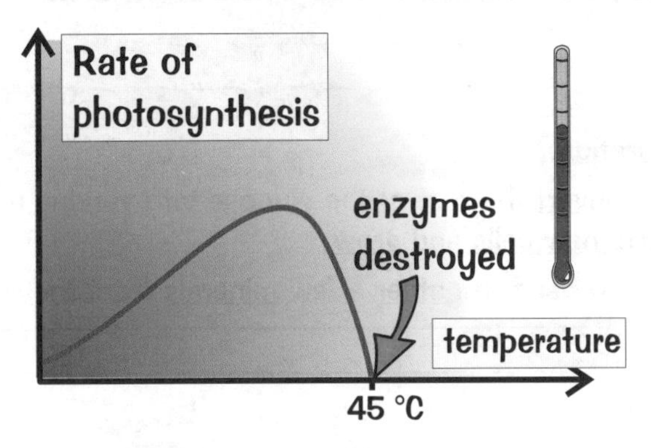

1) Usually, if the temperature is the limiting factor it's because it's too low — the enzymes needed for photosynthesis work more slowly at low temperatures.
2) But if the plant gets too hot, the enzymes it needs for photosynthesis and its other reactions will be damaged (see p.53).
3) This happens at about 45 °C (which is pretty hot for outdoors, although greenhouses can get that hot if you're not careful).
4) Experimentally, the best way to control the temperature of the flask is to put it in a water bath.

In all these experiments, you have to try and keep all the variables constant apart from the one you're investigating, so it's a fair test:

- use a bench lamp to control the intensity of the light (careful not to block the light with anything)
- keep the flask in a water bath to help keep the temperature constant
- you can't really do anything about the CO_2 levels — you just have to use a large flask, and do the experiments as quickly as you can, so that the plant doesn't use up too much of the CO_2 in the flask. If you're using sodium hydrogencarbonate make sure it's changed each time.

You can Artificially Create the Ideal Conditions for Farming

1) The most common way to artificially create the ideal environment for plants is to grow them in a greenhouse.
2) Greenhouses help to trap the sun's heat, and make sure that the temperature doesn't become limiting. In winter a farmer or gardener might use a heater as well to keep the temperature at the ideal level. In summer it could get too hot, so they might use shades and ventilation to cool things down.
3) Light's always needed for photosynthesis, so commercial farmers often supply artificial light after the Sun goes down to give their plants more quality photosynthesis time.

4) Farmers and gardeners can also increase the level of carbon dioxide in the greenhouse. A fairly common way is to use a paraffin heater to heat the greenhouse. As the paraffin burns, it makes carbon dioxide as a by-product.
5) Keeping plants enclosed in a greenhouse also makes it easier to keep them free from pests and diseases. The farmer can add fertilisers to the soil as well, to provide all the minerals needed for healthy growth (see page 47).
6) If the farmer can keep the conditions just right for photosynthesis, the plants will grow much faster and a decent crop can be harvested much more often.

Don't blame it on the sunshine, don't blame it on the CO_2...

...don't blame it on the temperature, blame it on the plant. Right, and now you'll never forget the three limiting factors in photosynthesis. No... well, make sure you read these pages over and over again till you do. With your newly found knowledge of photosynthesis you could take over the world...

How Plants Use the Glucose

Once plants have made the glucose, there are various ways they can use it.

① For Respiration

1) Plants manufacture glucose in their leaves.
2) They then use some of the glucose for respiration.
3) This releases energy which enables them to convert the rest of the glucose into various other useful substances which they can use to build new cells and grow.
4) To produce some of these substances they also need to gather a few minerals from the soil.

② Making Fruits

Glucose, along with another sugar called fructose, is turned into sucrose for storing in fruits. Fruits deliberately taste nice so that animals will eat them and spread the seeds all over the place in their poo.

③ Making Cell Walls

Glucose is converted into cellulose for making cell walls, especially in a rapidly growing plant.

④ Making Proteins

Glucose is combined with nitrates (collected from the soil) to make amino acids, which are then made into proteins.

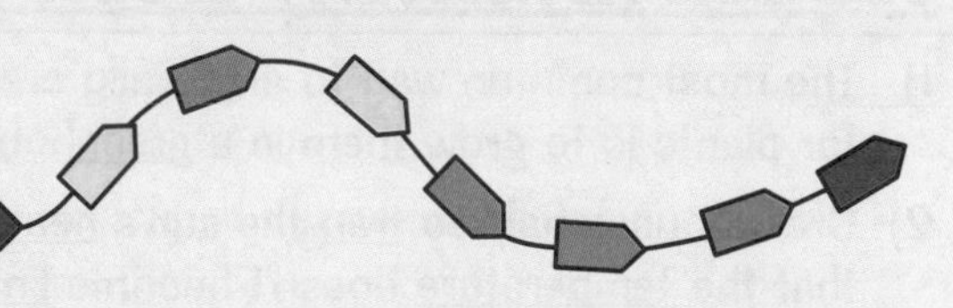

⑤ Stored in Seeds

Glucose is turned into lipids (fats and oils) for storing in seeds. Sunflower seeds, for example, contain a lot of oil — we get cooking oil and margarine from them. Seeds also store starch (see below).

⑥ Stored as Starch

Glucose is turned into starch and stored in roots, stems and leaves, ready for use when photosynthesis isn't happening, like in the winter.

Starch is insoluble which makes it much better for storing, because it doesn't bloat the storage cells by osmosis like glucose would.

Potato and carrot plants store a lot of starch underground over the winter so a new plant can grow from it the following spring. We eat the swollen storage organs.

Or for making small ornamental birdcages...

Actually, I made that last one up. I was bored. So there are actually only six things to learn that plants do with glucose. Right, shut the book right now. Or actually, finish reading this and then shut the book. Then write down all six uses of glucose from memory. Bet you forget one. Repeat until you don't.

Minerals for Healthy Growth

Plants need various mineral salts, as well as the carbohydrates they make by photosynthesis, in order to grow properly. They get these mineral ions from the soil by absorbing them through their roots:

You Need to Know About Two Minerals in Particular

1) Nitrates

Nitrates are needed for making amino acids, which are then used to make proteins.

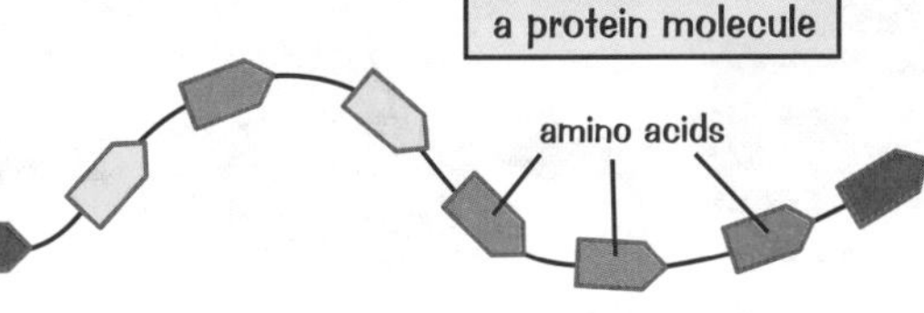

2) Magnesium

Magnesium is needed to make chlorophyll, which in turn is needed for photosynthesis.

Other minerals needed by plants include potassium and phosphates, which are used for things like making DNA and cell membranes, and helping the enzymes involved in photosynthesis and respiration to work properly.

Lack of These Nutrients Causes Deficiency Symptoms

Sometimes plants can't get all of the mineral ions they need to be healthy. It depends what's there in the soil — if the supply of nitrates in the soil gets low, the plant can't just wander off and find some more. It has to put up with it, and eventually it will start to show symptoms of the deficiency.

1) Lack of Nitrates

If the soil is deficient in nitrates, the plant starts to show stunted growth and won't reach its usual size. This is because proteins are needed for new growth, and they can't be made without nitrates.

2) Lack of Magnesium

If the soil is deficient in magnesium, the leaves of the plant start to turn yellow. This is because magnesium is needed to make chlorophyll, and this gives leaves their green colour.

If the plant is left short of the minerals it needs for a long time, it might die.

Deficiencies Can be Caused by Monoculture

1) Monoculture is where just one type of crop is grown in the same field year after year.
2) All the plants are the same crop, so they need the same minerals. This means the soil becomes deficient in the minerals which that crop uses lots of.
3) Deficiency of just one mineral is enough to cause poor growth and give a reduced yield.
4) This soon results in poor crops unless fertiliser is added to replenish the depleted minerals.

Just relax and absorb the information...

When a farmer or a gardener buys fertiliser, that's pretty much what they're buying — a nice big bag of mineral salts to provide all the extra elements plants need to grow. The one they usually need most of is nitrate, which is why manure works quite well — it's full of nitrogenous waste excreted by animals.

Pyramids of Number and Biomass

A trophic level is a feeding level. It comes from the Greek word trophe meaning 'nourishment'. So there.

You Need to Be Able to Construct Pyramids of Biomass

There's less energy and less biomass every time you move up a stage (trophic level) in a food chain. There are usually fewer organisms every time you move up a level too:

100 dandelions... feed... 10 rabbits... which feed... one fox.

This isn't always true though — for example, if 500 fleas are feeding on the fox, the number of organisms has increased as you move up that stage in the food chain. So a better way to look at the food chain is often to think about biomass instead of number of organisms. You can use information about biomass to construct a pyramid of biomass to represent the food chain:

1) Each bar on a pyramid of biomass shows the mass of living material at that stage of the food chain — basically how much all the organisms at each level would "weigh" if you put them all together.
2) So the one fox above would have a big biomass and the hundreds of fleas would have a very small biomass. Biomass pyramids are practically always the right shape (unlike number pyramids):

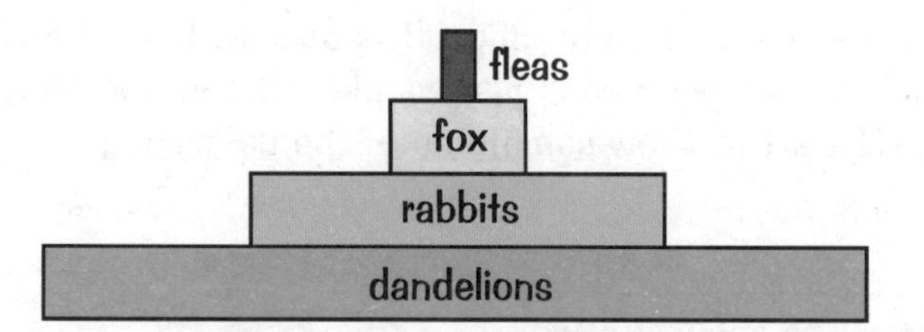

You need to be able to construct pyramids of biomass. Luckily it's pretty simple — they'll give you all the information you need to do it in the exam.

The big bar along the bottom of the pyramid always represents the producer (i.e. a plant). The next bar will be the primary consumer (the animal that eats the plant), then the secondary consumer (the animal that eats the primary consumer) and so on up the food chain. Easy.

You Need to be Able to Interpret Pyramids of Biomass

You also need to be able to look at pyramids of biomass and explain what they show about the food chain. Also very easy. For example:

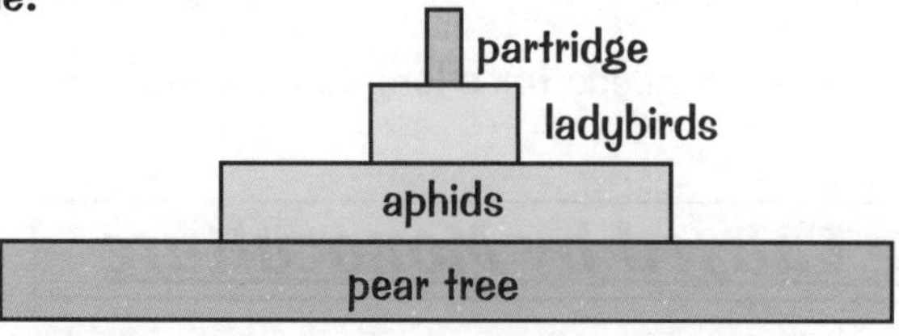

Even if you know nothing about the natural world, you're probably aware that a tree is quite a bit bigger than an aphid. So what's going on here is that lots (probably thousands) of aphids are feeding on a few great big trees. Quite a lot of ladybirds are then eating the aphids, and a few partridges are eating the ladybirds. Biomass and energy are still decreasing as you go up the levels — it's just that one tree can have a very big biomass, and can fix a lot of the Sun's energy using all those leaves.

Constructing pyramids is a breeze — just ask the Egyptians...

There are actually a couple of exceptions where pyramids of biomass aren't quite pyramid-shaped. It happens when the producer has a very short life but reproduces loads, like with plankton at certain times of year. But it's rare, and you don't need to know about it. Forget I ever mentioned it. Sorry.

Energy Transfer and Decay

So now you need to learn why there's less energy and biomass every time you move up a level.

All That Energy Just Disappears Somehow...

1) Energy from the Sun is the source of energy for nearly all life on Earth.

2) Plants use a small percentage of the light energy from the Sun to make food during photosynthesis. This energy's stored in the substances which make up the cells of plants, and then works its way through the food web as animals eat the plants and each other.

3) Respiration (see page 54), supplies the power for all life processes, including movement. Most of the energy is eventually lost to the surroundings as heat. This is especially true for mammals and birds, whose bodies must be kept at a constant temperature which is normally higher than their surroundings.

Material and energy are both lost at each stage of the food chain.

HEAT LOSS

MATERIALS LOST IN ANIMAL'S WASTE

4) Some of the material which makes up plants and animals is inedible (e.g. bone), so it doesn't pass to the next stage of the food chain. Material and energy are also lost from the food chain in the droppings — excretion.

This explains why you get biomass pyramids. Most of the biomass is lost and so does not become biomass in the next level up.

(There's more about the energy stored in biomass on page 48.)

It also explains why you hardly ever get food chains with more than about five trophic levels. So much energy is lost at each stage that there's not enough left to support more organisms after four or five stages.

Elements are Cycled Back to the Start of the Food Chain by Decay

1) Living things are made of materials they take from the world around them.
2) Plants take elements like carbon, oxygen, hydrogen and nitrogen from the soil or the air. They turn these elements into the complex compounds (carbohydrates, proteins and fats) that make up living organisms, and these then pass through the food chain.
3) These elements are returned to the environment in waste products produced by the organisms, or when the organisms die. These materials decay because they're broken down (digested) by microorganisms — that's how the elements get put back into the soil.
4) Microorganisms work best in warm, moist conditions. Many microorganisms also break down material faster when there's plenty of oxygen available.
5) All the important elements are thus recycled — they return to the soil, ready to be used by new plants and put back into the food chain again.
6) In a stable community the materials taken out of the soil and used are balanced by those that are put back in. There's a constant cycle happening.

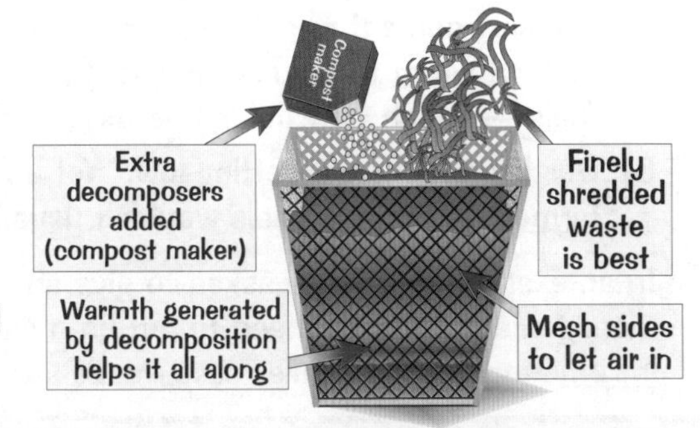

So when revising, put the fire on and don't take toilet breaks...

No, I'm being silly, go if you have to. We're talking in general terms about whole food chains here — you won't lose your concentration as a direct result of, erm, excretion.

Managing Food Production

People have been able to use what they know about energy loss from food chains to find the most efficient ways of producing food. But most efficient isn't necessarily best. Although it is often cheapest.

The "Efficiency" of Food Production Can Be Improved...

There are two ways to improve the efficiency of food production:

REDUCE THE NUMBER OF STAGES IN THE FOOD CHAIN

1) For a given area of land, you can produce a lot more food (for humans) by growing crops rather than by having grazing animals. This is because you are reducing the number of stages in the food chain. Only 10% of what beef cattle eat becomes useful meat for people to eat.
2) However, people do need to eat a varied diet to stay healthy, and there's still a lot of demand for meat products. Also remember that some land is unsuitable for growing crops, e.g. moorland or fellsides. In these places, animals like sheep and deer might be the best way to get food from the land.

RESTRICT THE ENERGY LOST BY FARM ANIMALS

1) In 'civilised' countries like the UK, animals such as pigs and chickens are often intensively farmed. They're kept close together indoors in small pens, so that they're warm and can't move about.
2) This saves them wasting energy on movement, and stops them giving out so much energy as heat. This makes the transfer of energy from the animal feed to the animal more efficient — so basically, the animals will grow faster on less food.
3) This makes things cheaper for the farmer, and for us when the animals finally turn up on supermarket shelves.

...but it Involves Compromises and Conflict

Improving the efficiency of food production is useful — it means cheaper food for us, and better standards of living for farmers. But it all comes at a cost.

1) Some people think that forcing animals to live in unnatural and uncomfortable conditions is cruel. There's a growing demand for organic meat, which means the animals will not have been intensively farmed.
2) The crowded conditions on factory farms create a favourable environment for the spread of diseases, like avian flu and foot-and-mouth disease.
3) To try to prevent disease, animals are given antibiotics. When the animals are eaten these can enter humans. This allows microbes that infect humans to develop immunity to those antibiotics — so the antibiotics become less effective as human medicines.
4) The environment where the animals are kept needs to be carefully controlled. The animals need to be kept warm to reduce the energy they lose as heat. This often means using power from fossil fuels — which we wouldn't be using if the animals were grazing in their natural environment.
5) Our fish stocks are getting low. Yet a lot of fish goes on feeding animals that are intensively farmed — these animals wouldn't usually eat this source of food.

In an exam, you may be asked to give an account of the positive and negative aspects of food management. You will need to put both sides, whatever your personal opinion is. If you get given some information on a particular case, make sure you use it — they want to see that you've read it carefully.

Locked in a little cage with no sunlight — who'd work in a bank...

You may well have quite a strong opinion on stuff like intensive farming of animals — whether it's 'tree-hugging hippie liberals, just give me a bit of nice cheap pork,' or 'poor creatures, they should be free, free as the wind!' Either way, keep it to yourself and give a nice, balanced argument instead.

The Carbon Cycle

As you've seen, all the nutrients in our environment are constantly being recycled — there's a nice balance between what goes in and what goes out again. This page is all about the recycling of carbon.

The Carbon Cycle Shows How Carbon is Recycled

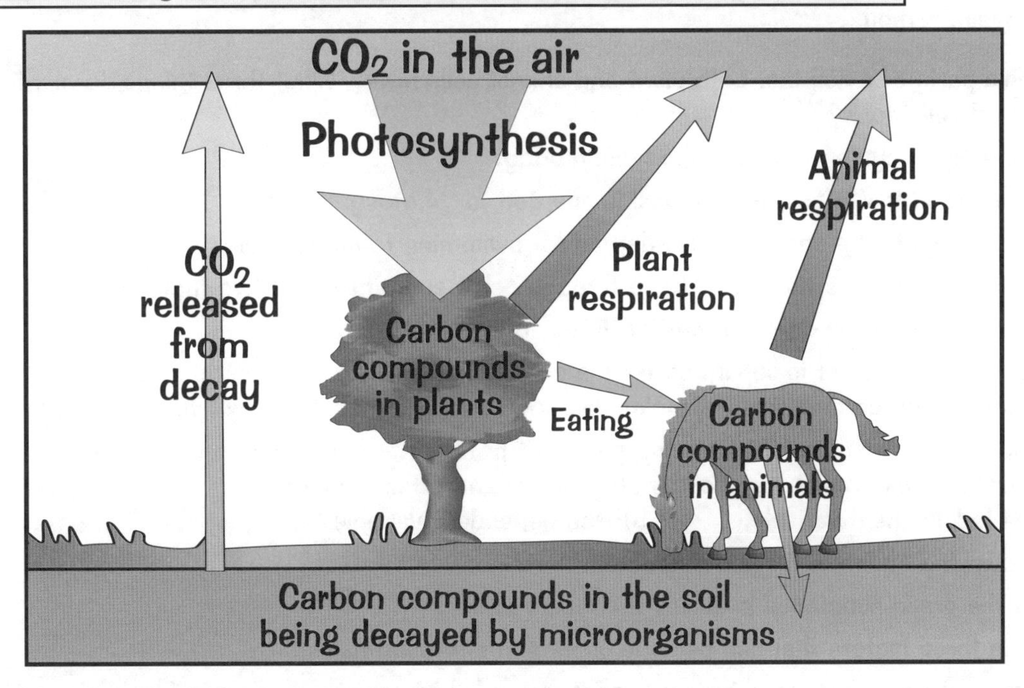

That can look a bit complicated at first, but it's actually pretty simple.
Learn these important points:

1) There's only one arrow going down from the atmosphere. The whole thing is "powered" by photosynthesis. CO_2 is removed from the atmosphere by green plants and used to make carbohydrates, fats and proteins in the plants.
2) Some of the CO_2 is returned to the atmosphere when the plants respire.
3) Some of the carbon becomes part of the compounds in animals when the plants are eaten. The carbon then moves through the food chain.
4) Some of the CO_2 is returned to the atmosphere when the animals respire.
5) When plants and animals die, other animals (called detritus feeders) and microorganisms feed on their remains. When these organisms respire, CO_2 is returned to the atmosphere.
6) Animals also produce waste, and this too is broken down by detritus feeders and microorganisms.
7) So the carbon is constantly being cycled — from the air, through food chains and eventually back out into the air again.

Carbon is also released into the atmosphere as CO_2 when plant and animal products are burnt.

What goes around comes around...

And that's the end of the section. But if you were hoping for a nice juicy bit of physics next (as if), or a tasty morsel of chemistry (yeah, right), I'm afraid you're going to be sadly disappointed. Yep, it's more of the same — biology. Oh come on, it could be worse. It could be vomit studies. Or poo analysis.

Revision Summary for Biology 2(i)

And where do you think you're going? It's no use just reading through and thinking you've got it all — this stuff will only stick in your head if you've learnt it properly. And that's what these questions are for. I won't pretend they'll be easy — they're not meant to be, but all the information's in the section somewhere. Have a go at all the questions, then if there are any you can't answer, go back, look stuff up and try again. Enjoy...

1) Name five parts of a cell that both plant and animal cells have. What three things do plant cells have that animal cells don't?
2) Name one organ system found in the human body.
3) Give three ways that a palisade leaf cell is adapted for photosynthesis.
4) Give three ways that a sperm cell is adapted for swimming to an egg cell.
5) Name three substances that can diffuse through cell membranes, and two that can't.
6) What three main things does the rate of diffusion depend on?
7) A solution of pure water is separated from a concentrated sugar solution by a partially permeable membrane. In which direction will molecules flow, and what substance will these molecules be?
8) An osmosis experiment involves placing pieces of potato into sugar solutions of various concentrations and measuring their lengths before and after. What is:
   a) the independent variable, b) the dependent variable?
9) Write down the equation for photosynthesis.
10) What is the green substance in leaves that absorbs sunlight?
11) Name the three factors that can become limiting in photosynthesis.
12) You carry out an experiment where you change the light intensity experienced by a piece of Canadian pondweed by changing the distance between the pondweed and a lamp supplying it with light. Write down four things which must be kept constant for this experiment to be a fair test.
13) Explain why it's important that a plant doesn't get too hot.
14) Describe three things that a gardener could do to make sure she gets a good crop of tomatoes.
15) Write down five ways that plants can use the glucose produced by photosynthesis.
16) Why is glucose turned into starch when plants need to store it for later?
17) What is the mineral magnesium needed for in a plant?
18) Describe the symptoms of nitrogen deficiency in a plant.
19) Why are farmers more likely to need extra fertiliser if they grow their crops as a monoculture?
20) What is meant by the term 'biomass'?
21) One oak tree produces acorns that are eaten by ten squirrels. At which stage in this section of the food chain is there the greatest:
   a) biomass, b) energy?
22) Give two ways that energy is lost from a food chain.
23) Explain why mammals and birds tend to lose more energy as heat than reptiles or insects.
24) Why do dead animals and plants decay after they die?
25) A farmer has a field. He plans to grow corn in it and then feed the corn to his cows, which he raises for meat. How could the farmer use the field more efficiently to produce food for humans?
26) Why do chickens kept in tiny cages in heated sheds need less food?
27) Summarise the main arguments for and against the intensive farming of animals.
28) Give one way that carbon dioxide from the air enters a food chain.
29) Give three ways that carbon compounds in a food chain become carbon dioxide in the air again.

Biological Catalysts — Enzymes

Chemical reactions are what make you work. And enzymes are what make them work.

Enzymes Are Catalysts Produced by Living Things

1) Living things have thousands of different chemical reactions going on inside them all the time.
2) These reactions need to be carefully controlled — to get the right amounts of substances.
3) You can usually make a reaction happen more quickly by raising the temperature. This would speed up the useful reactions but also the unwanted ones too... not good. There's also a limit to how far you can raise the temperature inside a living creature before its cells start getting damaged.
4) So... living things produce enzymes which act as biological catalysts. Enzymes reduce the need for high temperatures and we only have enzymes to speed up the useful chemical reactions in the body.

A CATALYST is a substance which INCREASES the speed of a reaction, without being CHANGED or USED UP in the reaction.

5) Enzymes are all proteins, which is one reason why proteins are so important to living things.
6) All proteins are made up of chains of amino acids. These chains are folded into unique shapes, which enzymes need to do their jobs (see below).

Enzymes Have Special Shapes So They Can Catalyse Reactions

1) Chemical reactions usually involve things either being split apart or joined together.
2) Every enzyme has a unique shape that fits onto the substance involved in a reaction.
3) Enzymes are really picky — they usually only catalyse one reaction.
4) This is because, for the enzyme to work, the substance has to fit its special shape. If the substance doesn't match the enzyme's shape, then the reaction won't be catalysed.

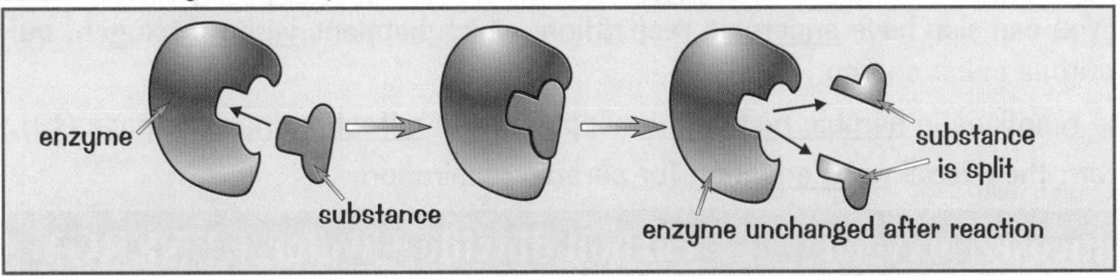

Enzymes Need the Right Temperature and pH

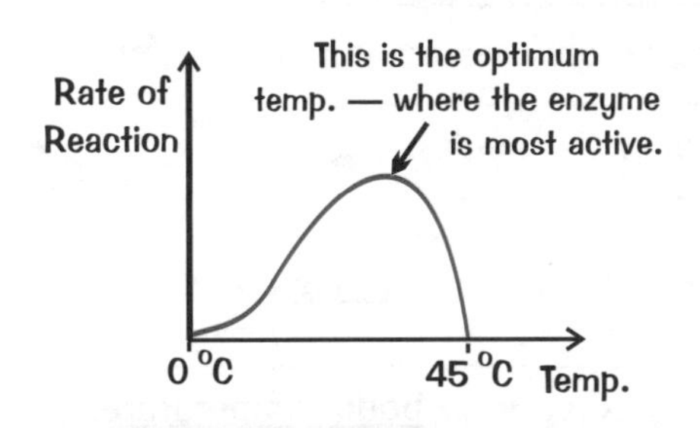

1) Changing the temperature changes the rate of an enzyme-catalysed reaction.
2) Like with any reaction, a higher temperature increases the rate at first. But if it gets too hot, some of the bonds holding the enzyme together break. This destroys the enzyme's special shape and so it won't work any more. It's said to be denatured.
3) Enzymes in the human body normally work best at around 37 °C — body temperature.
4) The pH also affects enzymes. If it's too high or too low, the pH interferes with the bonds holding the enzyme together. This changes the shape and denatures the enzyme.
5) All enzymes have an optimum pH that they work best at. It's often neutral pH 7, but not always — e.g. pepsin is an enzyme used to break down proteins in the stomach. It works best at pH 2, which means it's well-suited to the acidic conditions there.

Rate of reaction
Optimum pH
pH

If only enzymes could speed up revision...

Just like you've got to have the correct key for a lock, you've got to have the right substance for an enzyme. If the substance doesn't fit, the enzyme won't catalyse the reaction...

Enzymes and Respiration

Many chemical reactions inside cells are controlled by enzymes — including the ones in respiration, protein synthesis and photosynthesis (see page 43).

Enzymes Help Build Amino Acids and Proteins

Enzymes are used to synthesise molecules like amino acids — the ones you don't get from your diet. They also catalyse protein synthesis by joining together amino acids. These proteins could be enzymes — so it all works in a bit of a circle really.

Respiration is NOT "Breathing In and Out"

Respiration involves many reactions, all of which are catalysed by enzymes. These are really important reactions, as respiration releases the energy that the cell needs to do just about everything.

1) Respiration is not breathing in and breathing out, as you might think.
2) Respiration is the process of releasing energy from the breakdown of glucose — and goes on in every cell in your body.
3) It happens in plants too. All living things respire. It's how they release energy from their food.

RESPIRATION is the process of RELEASING ENERGY FROM GLUCOSE, which goes on IN EVERY CELL.

Aerobic Respiration Needs Plenty of Oxygen

1) Aerobic respiration is respiration using oxygen. It's the most efficient way to release energy from glucose. (You can also have anaerobic respiration, which happens without oxygen, but that doesn't release nearly as much energy.)
2) Most of the reactions in aerobic respiration happen inside mitochondria (see page 39).

You need to learn the overall word equation for aerobic respiration:

Glucose + oxygen → carbon dioxide + water + ENERGY

Respiration Releases Energy for All Kinds of Things

You need to learn these four examples of what the energy released by aerobic respiration is used for:

1) To build up larger molecules from smaller ones (like proteins from amino acids).

2) In animals, to allow the muscles to contract (which in turn allows them to move about).

3) In mammals and birds the energy is used to keep their body temperature steady (unlike other animals, mammals and birds are warm-blooded).

4) In plants, to build sugars, nitrates and other nutrients into amino acids, which are then built up into proteins.

Breathe, 2, 3, 4 — and release, 6, 7, 8...

So... respiration — that's a pretty important thing. Cyanide is a really nasty toxin that stops respiration by affecting enzymes involved in the process — so it's pretty poisonous (it can kill you). Your brain, heart and liver are affected first because they have the highest energy demands... nice.

Enzymes and Digestion

The enzymes used in respiration work inside cells. Various different enzymes are used in digestion too, but these enzymes are produced by specialised cells and then released into the gut to mix with the food.

Digestive Enzymes Break Down Big Molecules into Smaller Ones

1) Starch, proteins and fats are BIG molecules. They're too big to pass through the walls of the digestive system.
2) Sugars, amino acids, glycerol and fatty acids are much smaller molecules. They can pass easily through the walls of the digestive system.
3) The digestive enzymes break down the BIG molecules into the smaller ones.

Amylase Converts Starch into Simple Sugars

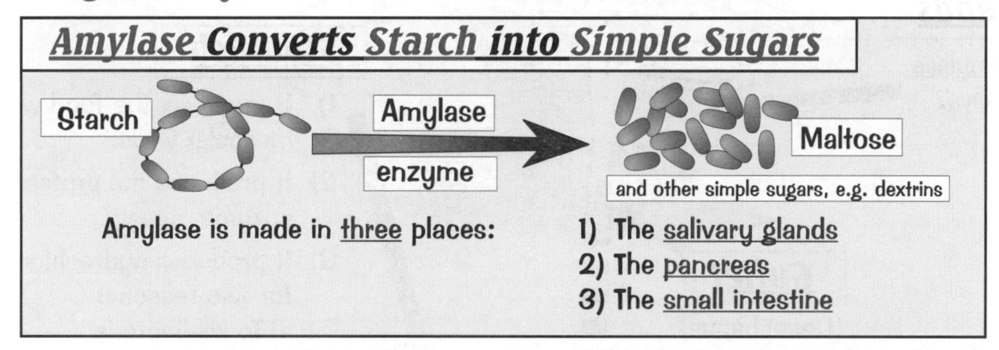

Amylase is made in three places:
1) The salivary glands
2) The pancreas
3) The small intestine

Protease Converts Proteins into Amino Acids

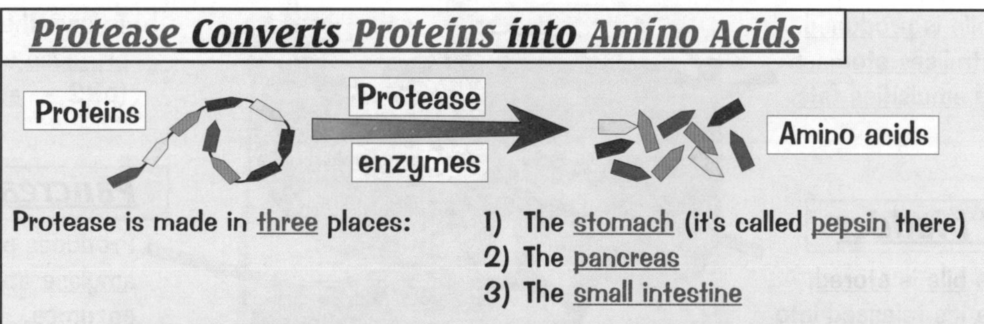

Protease is made in three places:
1) The stomach (it's called pepsin there)
2) The pancreas
3) The small intestine

Lipase Converts Fats into Glycerol and Fatty Acids

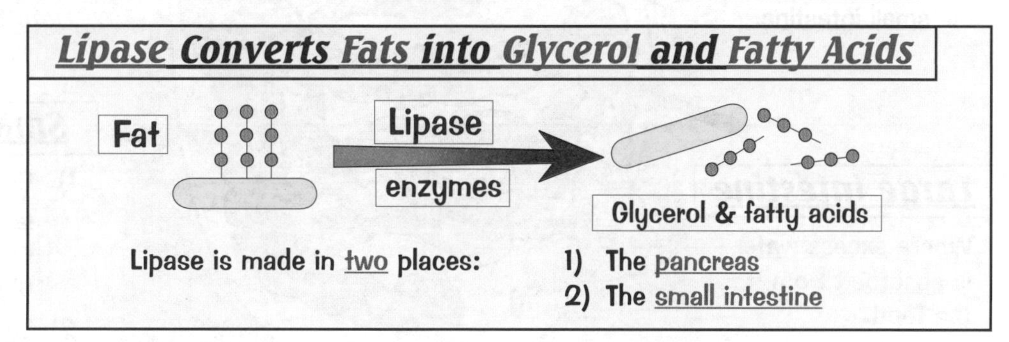

Lipase is made in two places:
1) The pancreas
2) The small intestine

Bile Neutralises the Stomach Acid and Emulsifies Fats

1) Bile is produced in the liver. It's stored in the gall bladder before it's released into the small intestine.
2) The hydrochloric acid in the stomach makes the pH too acidic for enzymes in the small intestine to work properly. Bile is alkaline — it neutralises the acid and makes conditions alkaline. The enzymes in the small intestine work best in these alkaline conditions.
3) It emulsifies fats. In other words it breaks the fat into tiny droplets. This gives a much bigger surface area of fat for the enzyme lipase to work on — which makes its digestion faster.

What do you call an acid that's eaten all the pies...

This all happens inside our digestive system, but there are some microorganisms which secrete their digestive enzymes outside their body onto the food. The food's digested, then the microorganism absorbs the nutrients. Nice. I wouldn't like to empty the contents of my stomach onto my plate before eating it.

The Digestive System

So now you know what the enzymes do, here's a nice big picture of the whole of the digestive system.

The Breakdown of Food is Catalysed by Enzymes

1) Enzymes used in the digestive system are produced by specialised cells in glands and in the gut lining.
2) Different enzymes catalyse the breakdown of different food molecules.

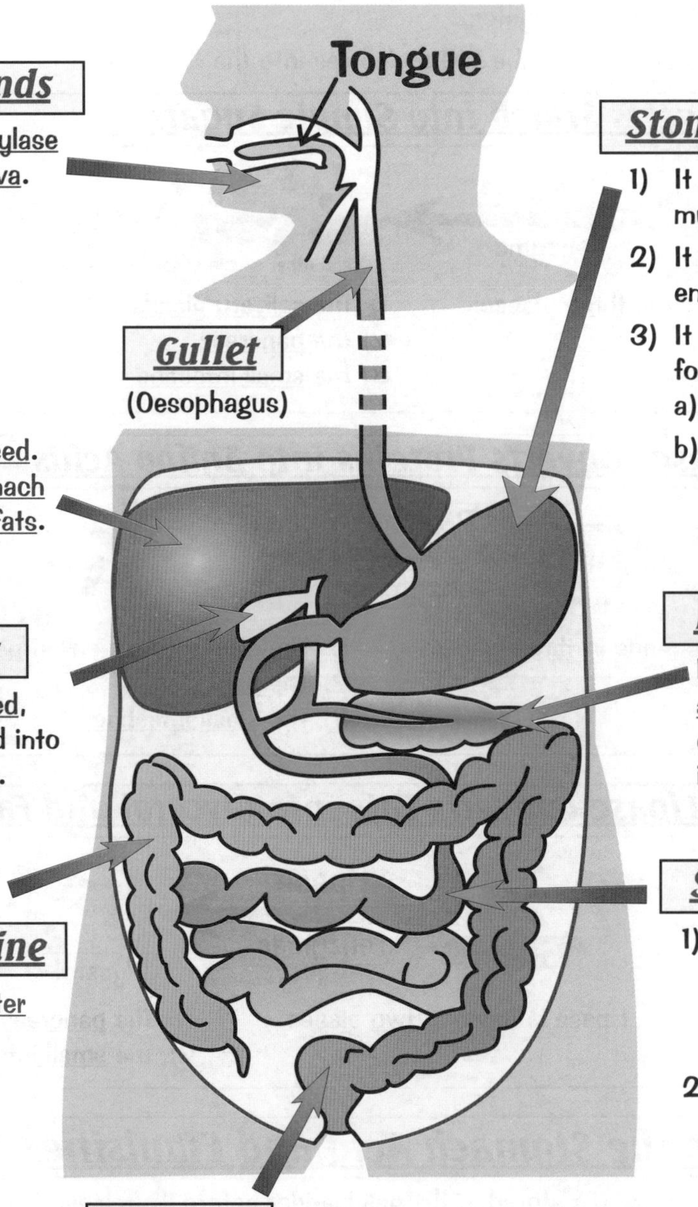

Tongue

These produce amylase enzyme in the saliva.

Stomach

1) It pummels the food with its muscular walls.
2) It produces the protease enzyme, pepsin.
3) It produces hydrochloric acid for two reasons:
   a) To kill bacteria
   b) To give the right pH for the protease enzyme to work (pH2 — acidic).

Gullet

(Oesophagus)

Liver

Where bile is produced. Bile neutralises stomach acid and emulsifies fats.

Gall bladder

Where bile is stored, before it's released into the small intestine.

Pancreas

Produces protease, amylase and lipase enzymes. It releases these into the small intestine.

Small intestine

1) Produces protease, amylase and lipase enzymes to complete digestion.
2) This is also where the "food" is absorbed out of the digestive system into the body.

Large intestine

Where excess water is absorbed from the food.

Rectum

Where the faeces (made up mainly of indigestible food) are stored before they bid you a fond farewell through the anus.

Mmmm — so who's for a chocolate digestive...

Did you know that the whole of your digestive system is actually a hole that goes right through your body. Think about it. It just gets loads of food, digestive juices and enzymes piled into it. Most of it's then absorbed back into the body and the rest is politely stored ready for removal.

Uses of Enzymes

Some microorganisms produce enzymes which pass out of their cells and catalyse reactions outside them (e.g. to digest the microorganism's food). These enzymes have many uses in the home and in industry.

Enzymes Are Used in Biological Detergents

1) Enzymes are the 'biological' ingredients in biological detergents and washing powders.
2) They're mainly protein-digesting enzymes (proteases) and fat-digesting enzymes (lipases).
3) Because the enzymes attack animal and plant matter, they're ideal for removing stains like food or blood.

Enzymes Are Used to Change Foods

1) The proteins in some baby foods are 'pre-digested' using protein-digesting enzymes (proteases), so they're easier for the baby to digest.
2) Carbohydrate-digesting enzymes (carbohydrases) can be used to turn starch syrup (yuk) into sugar syrup (yum).
3) Glucose syrup can be turned into fructose syrup using an isomerase enzyme. Fructose is sweeter, so you can use less of it — good for slimming foods and drinks.

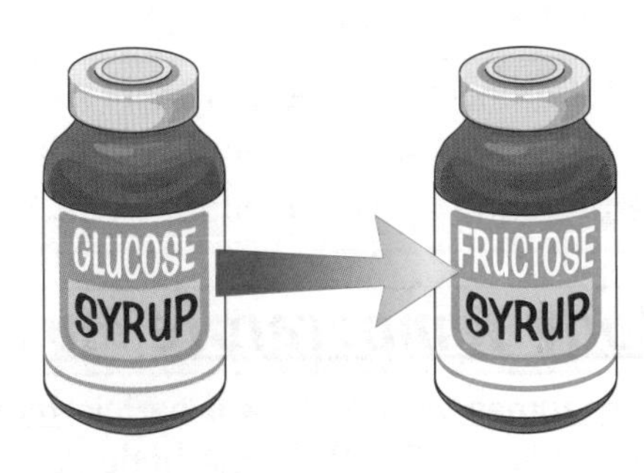

Using Enzymes in Industry Takes a Lot of Control

Enzymes are really useful in industry. They speed up reactions without the need for high temperatures and pressures. In a big industrial plant the substances are often continually run over the enzymes, so they have to be kept from washing away. They can be trapped in an alginate bead (a bead of jelly-like stuff) or in a latticework of silica gel.

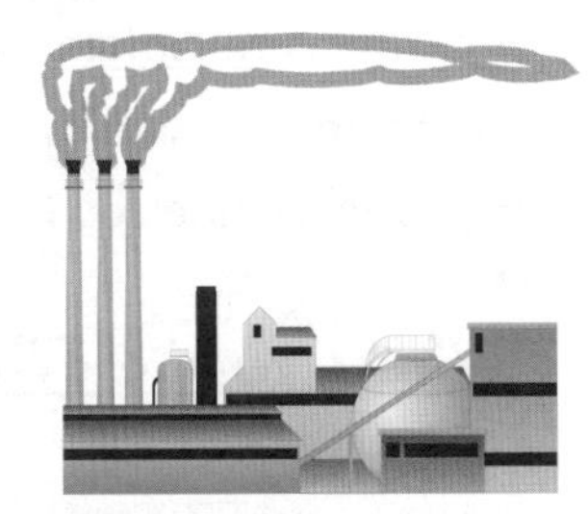

You need to know the advantages and disadvantages of using them, so here are a few to get you started:

ADVANTAGES

1) They're specific, so they only catalyse the reaction you want them to.
2) Using lower temperatures and pressures means a lower cost and it saves energy.
3) Enzymes work for a long time, so after the initial cost of buying them, you can continually use them.
4) They are biodegradable and therefore cause less environmental pollution.

DISADVANTAGES

1) Some people can develop allergies to the enzymes (e.g. in biological washing powders).
2) Enzymes can be denatured by even a small increase in temperature. They're also susceptible to poisons and changes in pH. This means the conditions in which they work must be tightly controlled.
3) Contamination of the enzyme with other substances can affect the reaction.

There's a lot to learn — but don't be deterred gents...

Enzymes are so picky. Even tiny little changes in pH or temperature will stop them working at maximum efficiency. They only catalyse one reaction as well, so you need to use a different one for each reaction. Temperamental little things these enzymes...

Homeostasis

Homeostasis is a fancy word. It covers lots of things, so I guess it has to be. Homeostasis covers all the functions of your body which try to maintain a "constant internal environment". Learn that definition:

HOMEOSTASIS is the maintenance of a constant internal environment.

There Are Six Main Things That Need to Be Controlled

The first four are all things you need, but at just the right level — not too much and not too little.

1) The body temperature can't get too hot or too cold (see below).
2) Water content mustn't get too high or low, or too much water could move into or out of cells and damage them. There's more on controlling water content on page 59.
3) If the ion content of the body is wrong, the same thing could happen. See page 59.
4) The blood sugar level needs to stay within certain limits (see pages 60 and 61).

The last two are waste products — they're constantly produced in the body and you need to get rid of them.

5) Carbon dioxide is a product of respiration. It's toxic in high quantities so it's got to be removed. It leaves the body by the lungs when you breathe out.
6) Urea is a waste product made from excess amino acids. There's more about it below.

Body Temperature Must Be Carefully Controlled

All enzymes work best at a certain temperature (see page 53). The enzymes within the human body work best at about 37 °C. If the body gets too hot or too cold, the enzymes won't work properly and some really important reactions could be disrupted. In extreme cases, this can even lead to death.

1) There is a thermoregulatory centre in the brain which acts as your own personal thermostat.
2) It contains receptors that are sensitive to the temperature of the blood flowing through the brain.
3) The thermoregulatory centre also receives impulses from the skin, giving info about skin temperature.
4) If you're getting too hot or too cold, your body can respond to try and cool you down or warm you up:

When you're TOO HOT:

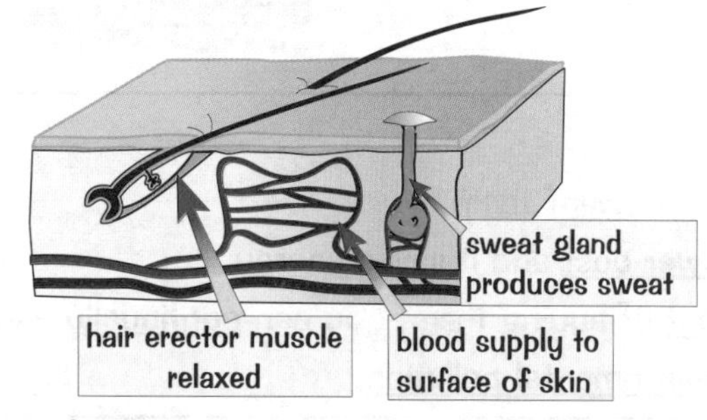

1) Hairs lie flat.
2) Sweat is produced by sweat glands and evaporates from the skin, which removes heat.
3) The blood vessels supplying the skin dilate so more blood flows close to the surface of the skin. This makes it easier for heat to be transferred from the blood to the environment.

When you're TOO COLD:

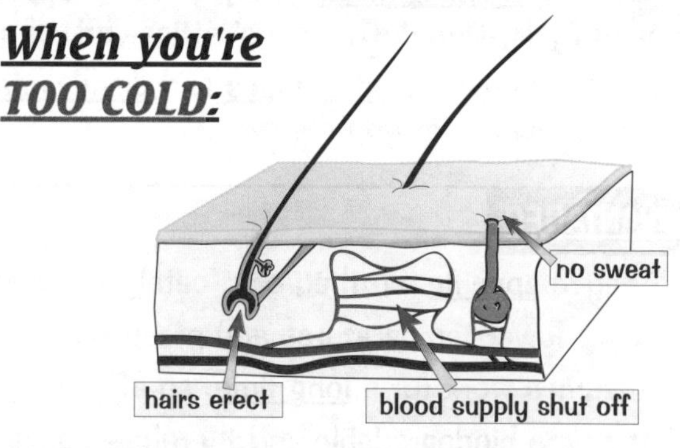

1) Hairs stand up to trap an insulating layer of air.
2) No sweat is produced.
3) Blood vessels supplying skin capillaries constrict to close off the skin's blood supply.

When you're cold you shiver too (your muscles contract automatically). This needs respiration, which releases some energy as heat.

Shiver me timbers — it's a wee bit nippy in here...

People who are exposed to extreme cold for long periods of time without protection can get frostbite — the blood supply to the fingers and toes is cut off to conserve heat (but this kills the cells, and they go black)... yuk

The Kidneys and Homeostasis

Kidneys are really important in this whole homeostasis thing.

Kidneys Basically Act as Filters to "Clean the Blood"

The kidneys perform three main roles:

1) Removal of urea from the blood.
2) Adjustment of ions in the blood.
3) Adjustment of water content of the blood.

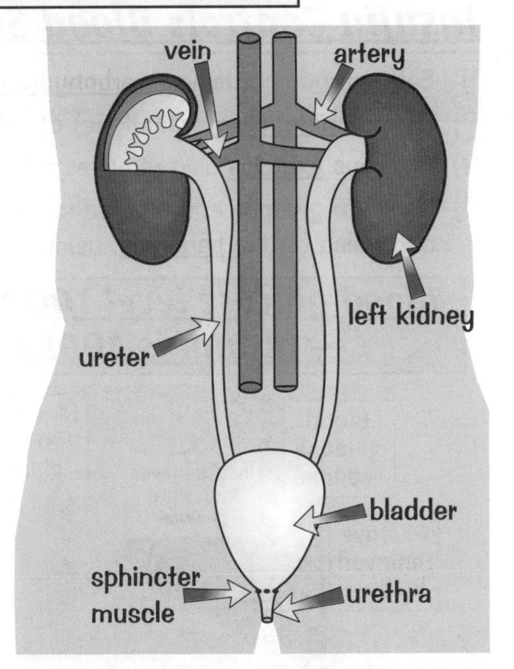

1) Removal of Urea

1) Proteins can't be stored by the body — so any excess amino acids are converted into fats and carbohydrates, which can be stored.
2) This process occurs in the liver. Urea is produced as a waste product from the reactions.
3) Urea is poisonous. It's released into the bloodstream by the liver. The kidneys then filter it out of the blood and it's excreted from the body in urine.

2) Adjustment of Ion Content

1) Ions such as sodium are taken into the body in food, and then absorbed into the blood.
2) If the ion content of the body is wrong, this could mean too much or too little water is drawn into cells by osmosis (see page 42). Having the wrong amount of water can damage cells.
3) Excess ions are removed by the kidneys. For example, a salty meal will contain far too much sodium and so the kidneys will remove the excess sodium ions from the blood.
4) Some ions are also lost in sweat (which tastes salty, you may have noticed).
5) But the important thing to remember is that the balance is always maintained by the kidneys.

3) Adjustment of Water Content

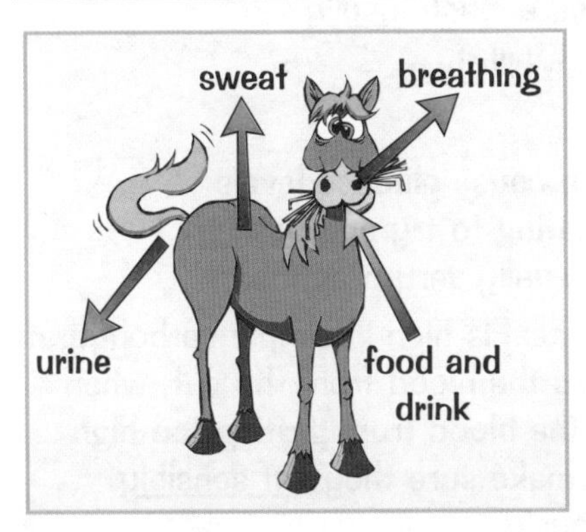

Water is taken into the body as food and drink and is lost from the body in three main ways:
1) In urine
2) In sweat
3) In the air we breathe out.

The body has to constantly balance the water coming in against the water going out. Our bodies can't control how much we lose in our breath, but we do control the other factors. This means the water balance is between:
1) Liquids consumed
2) Amount sweated out
3) Amount excreted by the kidneys in the urine.

On a cold day, if you don't sweat, you'll produce more urine which will be pale and dilute.
On a hot day, you sweat a lot, and you'll produce less urine which will be dark-coloured and concentrated.
The water lost when it is hot has to be replaced with water from food and drink to restore the balance.

Adjusting water content — blood, sweat and, erm, wee...

Scientists have made a machine which can do the kidney's job for us — a kidney dialysis machine. People with kidney failure have to use it for 3-4 hours, 3 times a week. Unfortunately it's not something you can carry around in your back pocket, which makes life difficult for people with kidney failure.

Controlling Blood Sugar

Blood sugar is also controlled as part of homeostasis. Insulin is a hormone that controls how much sugar there is in your blood. Learn how it does it:

Insulin Controls Blood Sugar Levels

1) Eating foods containing carbohydrate puts glucose into the blood from the gut.
2) Normal metabolism (reactions) of cells removes glucose from the blood.
3) Vigorous exercise also removes a lot of glucose from the blood.
4) Levels of glucose in the blood must be kept steady. Changes in blood glucose are monitored and controlled by the pancreas, using the hormone insulin, as shown:

Blood glucose level TOO HIGH — insulin is ADDED

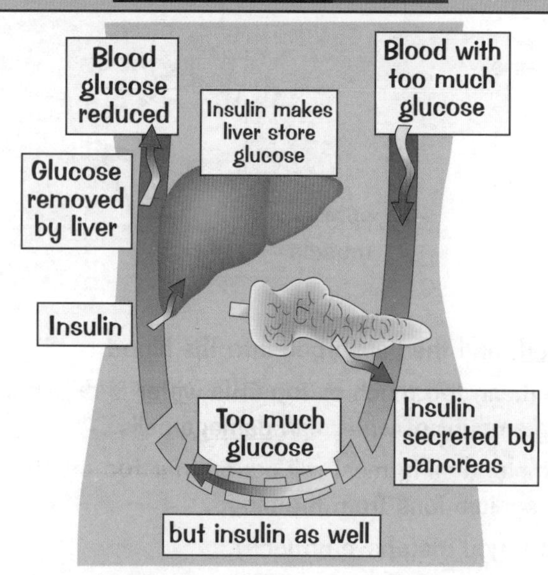

Blood glucose level TOO LOW — insulin is NOT ADDED

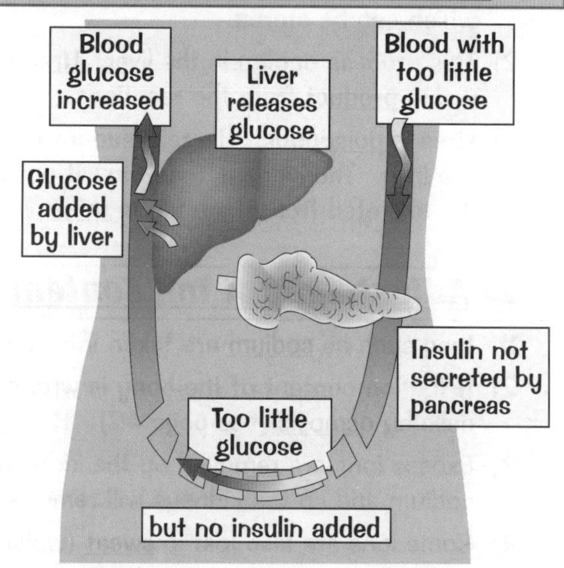

Diabetes (type 1) — the Pancreas Stops Making Enough Insulin

1) Diabetes (type 1) is a disorder where the pancreas doesn't produce enough insulin.
2) The result is that a person's blood sugar can rise to a level that can kill them.
3) The problem can be controlled in two ways:
   a) Avoiding foods rich in simple carbohydrates, i.e. sugars (which cause glucose levels to rise rapidly). It can also be helpful to take exercise after eating to try and use up the extra glucose produced during digestion — but this isn't usually very practical.
   b) Injecting insulin into the blood at mealtimes (especially if the meal is high in simple carbohydrates). This will make the liver remove the glucose as soon as it enters the blood from the gut, when the food is being digested. This stops the level of glucose in the blood from getting too high and is a very effective treatment. However, the person must make sure they eat sensibly after injecting insulin, or their blood sugar could drop dangerously.
4) The amount of insulin that needs to be injected depends on the person's diet and how active they are.
5) Diabetics can check their blood sugar using a glucose-monitoring device. This is a little hand-held machine. They prick their finger to get a drop of blood for the machine to check.

My blood sugar feels low after all that — pass the biscuits...

This stuff can seem a bit confusing at first, but if you concentrate on learning those two diagrams, it'll all start to get a lot easier. Don't forget that only carbohydrate foods put the blood sugar levels up.

Insulin and Diabetes

Scientific discoveries often take a long time, and a lot of trial and error — here's a rather famous example.

Insulin Was Discovered by Banting and Best

It has been known for some time that people who suffer from diabetes have a lot of sugar in their urine. In the 19th century, scientists removed pancreases from dogs, and the same sugary urine was observed — the dogs became diabetic. That suggested that the pancreas had to have something to do with the illness. In the 1920s Frederick Banting and his assistant Charles Best managed to successfully isolate insulin — the hormone that controls blood sugar levels.

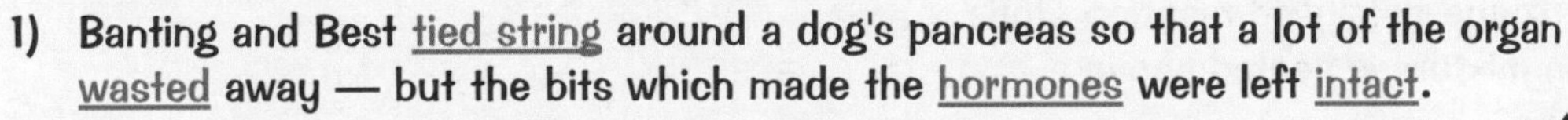

1) Banting and Best tied string around a dog's pancreas so that a lot of the organ wasted away — but the bits which made the hormones were left intact.
2) They removed the pancreas from the dog, and obtained an extract from it.
3) They then injected this extract into diabetic dogs and observed the effects on their blood sugar levels.
4) From the graph, you can see that after the pancreatic extract was injected, the dog's blood sugar level fell dramatically. This showed that the pancreatic extract caused a temporary decrease in blood sugar level.
5) They went on to isolate the substance in the pancreatic extract — insulin.

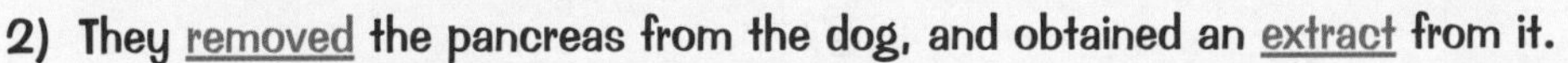

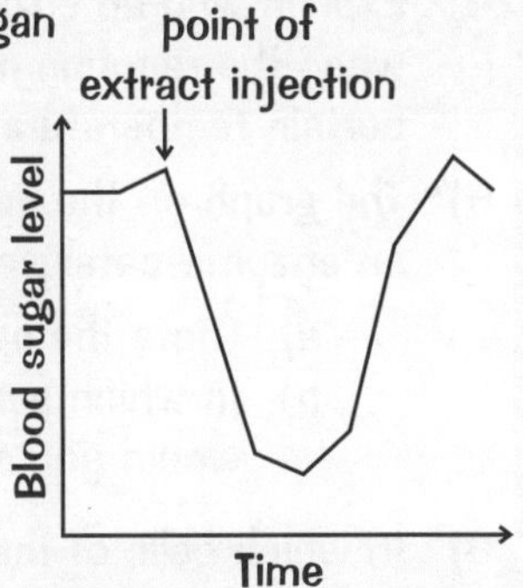

Diabetes Can Be Controlled by Regular Injections of Insulin

After a lot more experiments, Banting and Best tried injecting insulin into a diabetic human. And it worked. Since then insulin has been mass produced to meet the needs of diabetics. Diabetics have to inject themselves with insulin often — 2-4 times a day. They also need to carefully control their diet and the amount of exercise they do (see page 60).

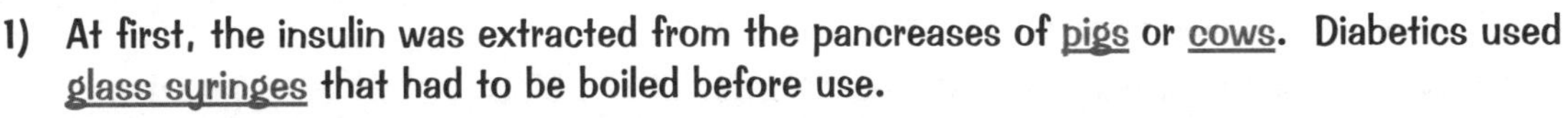

1) At first, the insulin was extracted from the pancreases of pigs or cows. Diabetics used glass syringes that had to be boiled before use.
2) In the 1980s human insulin made by genetic engineering became available. This didn't cause any adverse reactions in patients, which animal insulin sometimes did.

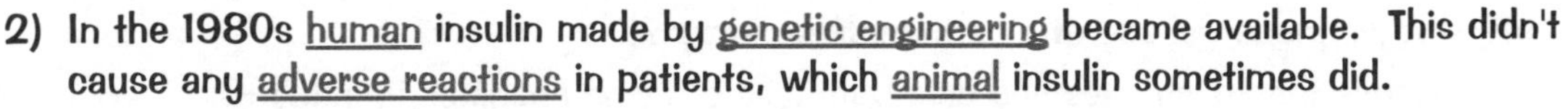

3) Slow, intermediate and fast acting insulins have been developed to make it easier for diabetics to control their blood sugar levels.
4) Ready sterilised, disposable syringes are now available, as well as needle-free devices.

Improving methods of treatment allow diabetics to control their blood sugar more easily. This helps them avoid some of the damaging side effects of poor control, such as blindness and gangrene.

Some Diabetics Are Offered a Pancreas Transplant

Injecting yourself with insulin every day controls the effects of diabetes, but it doesn't help to cure it.

1) Some diabetics are offered a pancreas transplant. A successful operation means they won't have to inject themselves with insulin again. But as with any organ transplant, your body can reject the tissue. This means you have to take costly immunosuppressive drugs, which often have serious side-effects.
2) Another method, still in its experimental stage, is to transplant just the cells which produce insulin. There's been varying success with this technique, and there are still problems with rejection.
3) Modern research into artificial pancreases and stem cell research may mean the elimination of organ rejection, but there's a way to go yet (see page 66).

Blimey — all that in the last hundred years...

Insulin can't be taken in a pill or tablet because the enzymes in the stomach completely destroy it before it reaches the bloodstream. That's the reason diabetics have to inject insulin.

Revision Summary for Biology 2(ii)

There are two quite separate bits to this section. First you've got enzymes, how they're used inside cells, in digestion, and in industry. Then the second bit is all about keeping things constant in your body. Have a bash at the questions, go back and check anything you're not sure about, then try again. Practise until you can answer all these questions really easily without having to look back at the section.

1) Give a definition of a catalyst.
2) Name three enzyme-catalysed chemical reactions that happen inside living organisms.
3) What family of molecules do enzymes belong to?
4) Explain why an enzyme-catalysed reaction stops when the reaction mixture is heated above a certain temperature.
5)* The graph on the right shows how the rate of an enzyme-catalysed reaction depends on pH:
    a) State the optimum pH of the enzyme.
    b) In which part of the human digestive system would you expect to find the enzyme?

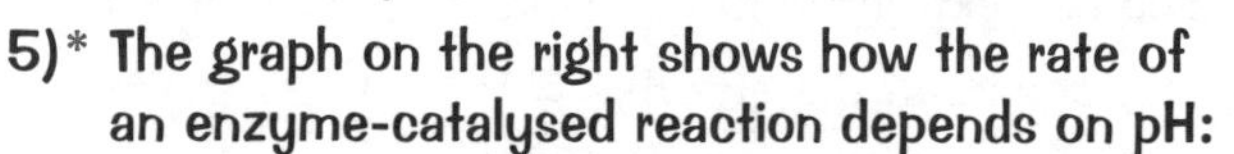
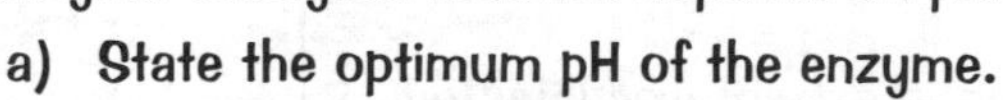
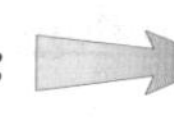
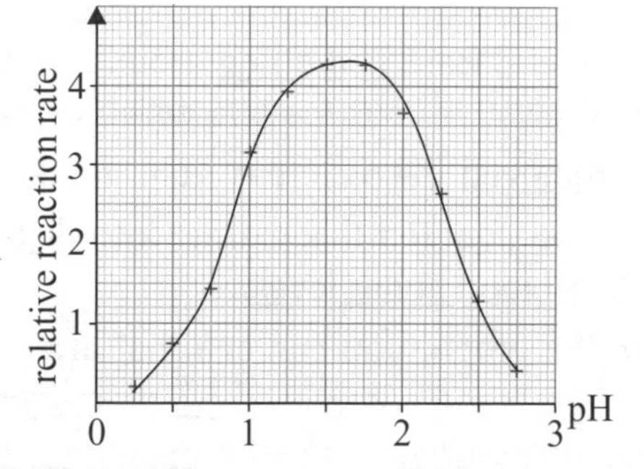

6) In which cells of the body does respiration happen? Where in the cell does respiration happen?
7) Write down the word equation for aerobic respiration.
8) Give three examples of life processes for which the energy from respiration is used:
    a) in a bird, b) in a plant.
9) In which three places in the body is amylase produced?
10) Where in the body is bile:
    a) produced? b) stored? c) used?
11) Explain why the stomach produces hydrochloric acid.
12) What is the main function of the small intestine?
13) Give two kinds of enzyme that would be useful in a biological washing powder.
14) Give an industrial use of a carbohydrase enzyme.
15) Discuss the advantages and disadvantages of using enzymes in industry.
16) Define homeostasis.
17) Write down four conditions that the body needs to keep fairly constant.
18) List two waste products that have to be removed from the body.
19) At what temperature do most of the enzymes in the human body work best?
20) Which area of the brain is involved in regulating the temperature of the body?
21) Write down three things that the body can do to reduce heat loss if it gets too cold.
22) What three main jobs do the kidneys do in the body?
23) Where in the body is urea produced?
24) What damage could be done in the body if the ion content is wrong?
25) Give three ways in which water is lost from the body.
26) Explain why your urine is likely to be more concentrated on a hot day.
27) Which organ monitors and controls blood glucose levels?
28) How does insulin lower the blood glucose level if it is too high?
29) How does the body respond if the blood glucose level is too low?
30) What causes diabetes? How is diabetes currently treated?
31) Describe the experiments by Banting and Best that led to the isolation of insulin.
32) Discuss the advantages and disadvantages of a pancreas transplant as a cure for diabetes.

* Answers on page 96

DNA

The first step in understanding genetics is getting to grips with DNA.

Chromosomes Are Really Long Molecules of DNA

1) DNA stands for deoxyribose nucleic acid.
2) It contains all the instructions to put an organism together and make it work.
3) It's found in the nucleus of animal and plant cells, in really long molecules called chromosomes.

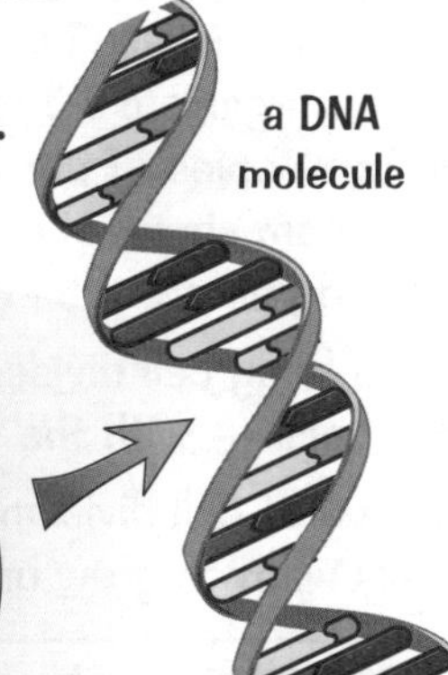

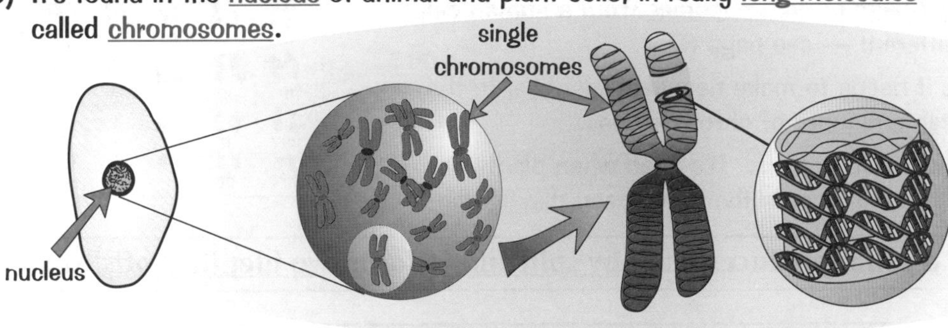

A Gene Codes for a Specific Protein

1) A gene is a section of DNA. It contains the instructions to make a specific protein.
2) Cells make proteins by stringing amino acids together in a particular order.
3) Only 20 amino acids are used, but they make up thousands of different proteins.
4) Genes simply tell cells in what order to put the amino acids together.
5) DNA also determines what proteins the cell produces, e.g. haemoglobin, keratin.
6) That in turn determines what type of cell it is, e.g. red blood cell, skin cell.

Everyone has Unique DNA...

...except identical twins and clones

Almost everyone's DNA is unique. The only exceptions are identical twins, where the two people have identical DNA, and clones.

DNA fingerprinting (or genetic fingerprinting) is a way of cutting up a person's DNA into small sections and then separating them. Every person's genetic fingerprint has a unique pattern (unless they're identical twins or clones of course). This means you can tell people apart by comparing samples of their DNA.

DNA fingerprinting is used in...

1) Forensic science — DNA (from hair, skin flakes, blood, semen etc.) taken from a crime scene is compared with a DNA sample taken from a suspect. In the diagram, suspect 1's DNA has the same pattern as the DNA from the crime scene — so suspect 1 was probably at the crime scene.
2) Paternity testing — to see if a man is the father of a particular child.

DNA from crime scene | suspect 1 | suspect 2 | suspect 3

Some people would like there to be a national genetic database of everyone in the country. That way, DNA from a crime scene could be checked against everyone in the country to see whose it was. But others think this is a big invasion of privacy, and they worry about how safe the data would be and what else it might be used for. There are also scientific problems — false positives can occur if errors are made in the procedure or if the data is misinterpreted.

So the trick is — frame your twin and they'll never get you...

In the exam you might have to interpret data on DNA fingerprinting for identification. They'd probably give you a diagram similar to the one at the bottom of this page, and you'd have to say which of the known samples (if any) matched the unknown sample. Pretty easy — it's the two that look the same.

Cell Division — Mitosis

In order to survive and grow, our cells have got to be able to divide. And that means our DNA as well...

Mitosis Makes New Cells for Growth and Repair

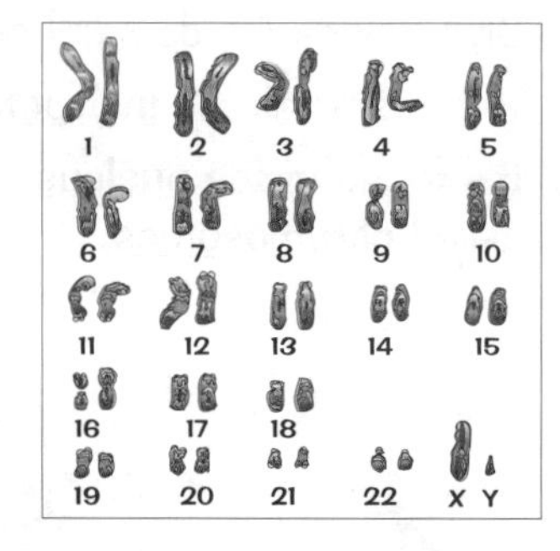

Body cells normally have two copies of each chromosome — one from the organism's 'mother', and one from its 'father'. So, humans have two copies of chromosome 1, two copies of chromosome 2, etc. The diagram shows the 23 pairs of chromosomes from a human cell. The 23rd pair are a bit different — see page 67.

When a body cell divides it needs to make new cells identical to the original cell — with the same number of chromosomes.

This type of cell division is called mitosis. It's used when plants and animals want to grow or to replace cells that have been damaged.

"**MITOSIS** is when a cell reproduces itself by splitting to form two identical offspring."

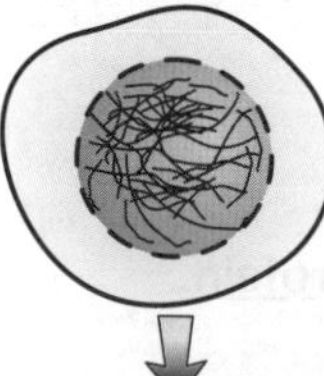

In a cell that's not dividing, the DNA is all spread out in long strings.

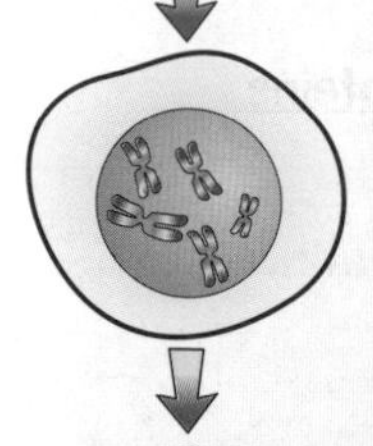

If the cell gets a signal to divide, it needs to duplicate its DNA — so there's one copy for each new cell. The DNA is copied and forms X-shaped chromosomes. Each 'arm' of the chromosome is an exact duplicate of the other.

The left arm has the same DNA as the right arm of the chromosome.

The chromosomes then line up at the centre of the cell and cell fibres pull them apart. The two arms of each chromosome go to opposite ends of the cell.

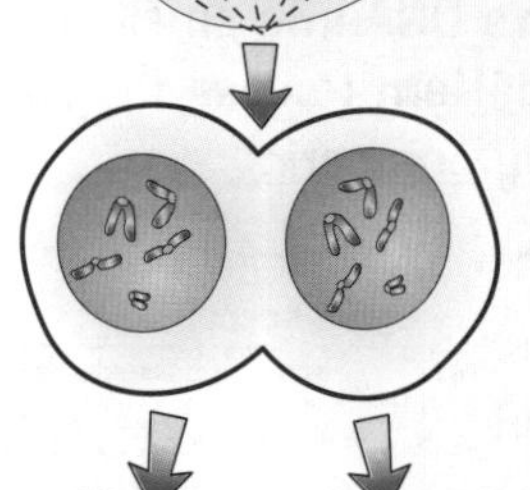

Membranes form around each of the sets of chromosomes. These become the nuclei of the two new cells.

Lastly, the cytoplasm divides.

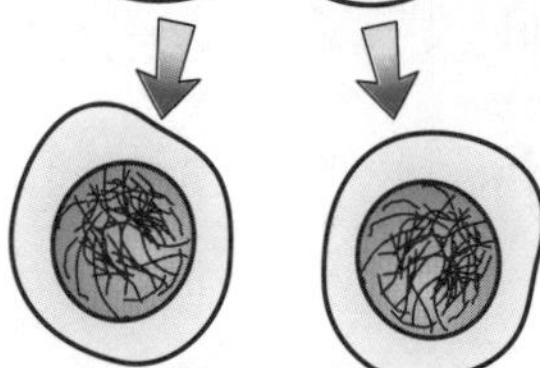

You now have two new cells containing exactly the same DNA — they're identical.

Asexual Reproduction Also Uses Mitosis

1) Some organisms also reproduce by mitosis, e.g. strawberry plants can form runners in this way, which become new plants.
2) This is an example of asexual reproduction.
3) The offspring have exactly the same genes as the parent — so there's no variation.

A cell's favourite computer game — divide and conquer...

This can seem tricky at first. But don't worry — just go through it slowly, one step at a time. This type of division produces identical cells, but there's another type which doesn't... (see next page)

Cells = 46 chromosomes Gametes = 23 chromosomes

Cell Division — Meiosis

You thought mitosis was exciting. Hah! You ain't seen nothing yet!

Gametes Have Half the Usual Number of Chromosomes

1) During sexual reproduction, two cells called gametes (sex cells) combine to form a new individual.
2) Gametes only have one copy of each chromosome. This is so that you can combine one sex cell from the 'mother' and one sex cell from the 'father' and still end up with the right number of chromosomes in body cells.
3) For example, human body cells have 46 chromosomes. The gametes have 23 chromosomes each, so that when an egg and sperm combine, you get 46 chromosomes again.

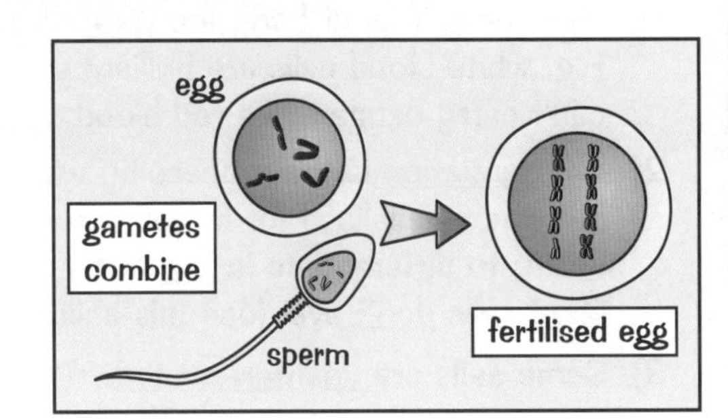

Meiosis Involves Two Divisions

To make new cells which only have half the original number of chromosomes, cells divide by meiosis. In humans it only happens in the reproductive organs (e.g. ovaries and testes).

"MEIOSIS produces cells which have half the normal number of chromosomes."

chromosome pair

As with mitosis, before the cell starts to divide, it duplicates its DNA — one arm of each chromosome is an exact copy of the other arm.

In the first division in meiosis (there are two divisions) the chromosome pairs line up in the centre of the cell.

The pairs are then pulled apart, so each new cell only has one copy of each chromosome. Some of the father's chromosomes (shown in blue) and some of the mother's chromosomes (shown in red) go into each new cell.

Each new cell will have a mixture of the mother's and father's chromosomes. Mixing up the genes in this way creates variation in the offspring. This is a huge advantage of sexual reproduction over asexual reproduction.

In the second division the chromosomes line up again in the centre of the cell. It's a lot like mitosis. The arms of the chromosomes are pulled apart.

You get four gametes each with only a single set of chromosomes in it.

After two gametes join at fertilisation, the cell grows by repeatedly dividing by mitosis.

Now that I have your undivided attention...

Remember, in humans, meiosis only occurs in reproductive organs, where gametes are being made.

Stem Cells

Stem cell research has exciting possibilities, but it's also pretty controversial.

Embryonic Stem Cells Can Turn into ANY Type of Cell

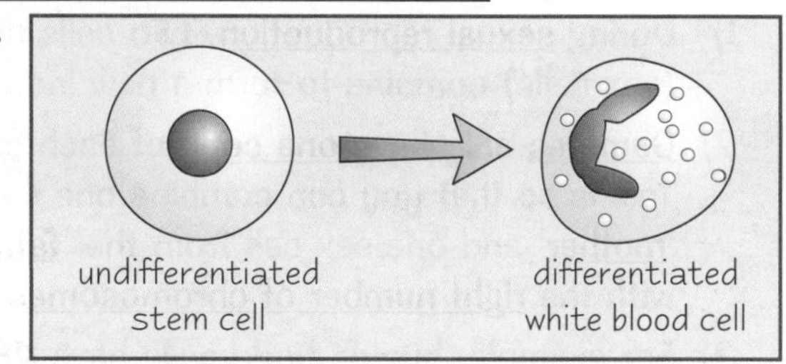

1) Most cells in your body are specialised for a particular job. E.g. white blood cells are brilliant at fighting invaders but can't carry oxygen, like red blood cells.
2) Differentiation is the process by which a cell changes to become specialised for its job. In most animal cells, the ability to differentiate is lost at an early stage, but lots of plant cells don't ever lose this ability.
3) Some cells are undifferentiated. They can develop into different types of cell depending on what instructions they're given. These cells are called STEM CELLS.
4) Stem cells are found in early human embryos. They're exciting to doctors and medical researchers because they have the potential to turn into any kind of cell at all. This makes sense if you think about it — all the different types of cell found in a human being have to come from those few cells in the early embryo.
5) Adults also have stem cells, but they're only found in certain places, like bone marrow. These aren't as versatile as embryonic stem cells — they can't turn into any cell type at all, only certain ones.

Stem Cells May Be Able to Cure Many Diseases

1) Medicine already uses adult stem cells to cure disease. For example, people with some blood diseases (e.g. sickle cell anaemia) can be treated by bone marrow transplants. Bone marrow contains stem cells that can turn into new blood cells to replace the faulty old ones.
2) Scientists can also extract stem cells from very early human embryos and grow them.

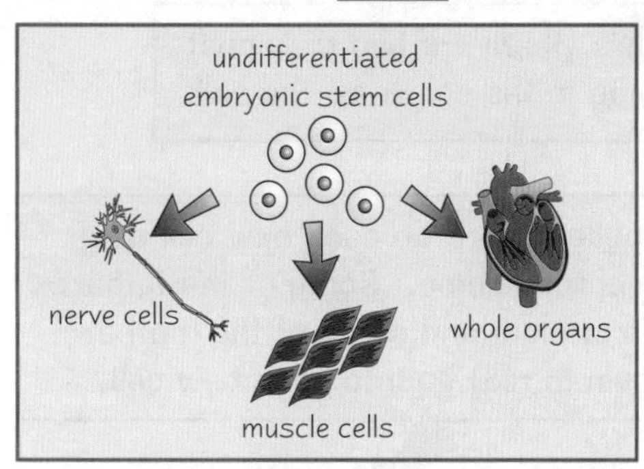

3) These embryonic stem cells could be used to replace faulty cells in sick people — you could make beating heart muscle cells for people with heart disease, insulin-producing cells for people with diabetes, nerve cells for people paralysed by spinal injuries, and so on.
4) To get cultures of one specific type of cell, researchers try to control the differentiation of the stem cells by changing the environment they're growing in. So far, it's still a bit hit and miss — lots more research is needed.

Some People Are Against Stem Cell Research

1) Some people are against stem cell research because they feel that human embryos shouldn't be used for experiments since each one is a potential human life.
2) Others think that curing patients who already exist and who are suffering is more important than the rights of embryos.
3) One fairly convincing argument in favour of this point of view is that the embryos used in the research are usually unwanted ones from fertility clinics which, if they weren't used for research, would probably just be destroyed. But of course, campaigners for the rights of embryos usually want this banned too.
4) These campaigners feel that scientists should concentrate more on finding and developing other sources of stem cells, so people could be helped without having to use embryos.
5) In some countries stem cell research is banned, but it's allowed in the UK as long as it follows strict guidelines.

But florists cell stems, and nobody complains about that...

The potential of stem cells is huge — but it's early days yet. Research has recently been done into getting stem cells from alternative sources. For example, some researchers think it might be possible to get cells from umbilical cords to behave like embryonic stem cells.

X and Y Chromosomes

Now for a couple of very important little chromosomes...

Your Chromosomes Control Whether You're Male or Female

There are 23 matched pairs of chromosomes in every human body cell. The 23rd pair are labelled XX or XY. They're the two chromosomes that decide whether you turn out male or female.

All men have an X and a Y chromosome: XY
The Y chromosome causes male characteristics.

All women have two X chromosomes: XX
The XX combination allows female characteristics to develop.

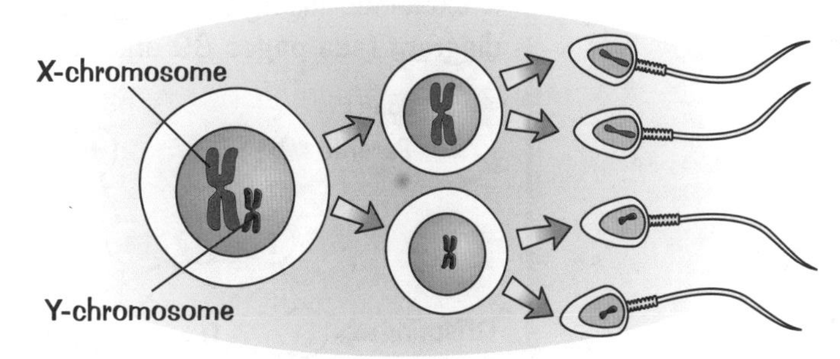

When making sperm, the X and Y chromosomes are drawn apart in the first division in meiosis. There's a 50% chance each sperm cell gets an X-chromosome and a 50% chance it gets a Y-chromosome.

A similar thing happens when making eggs. But the original cell has two X-chromosomes, so all the eggs have one X-chromosome.

Genetic Diagrams Show the Possible Combinations of Gametes

To find the probability of getting a boy or a girl, you can draw a genetic diagram.

Put the possible gametes from one parent down the side, and those from the other parent along the top. Then in each middle square you fill in the letters from the top and side that line up with that square. The pairs of letters in the middle show the possible combinations of the gametes.

There are two XX results and two XY results, so there's the same probability of getting a boy or a girl.

Don't forget that this 50:50 ratio is only a probability. If you had four kids they could all be boys — yes I know, terrifying isn't it?

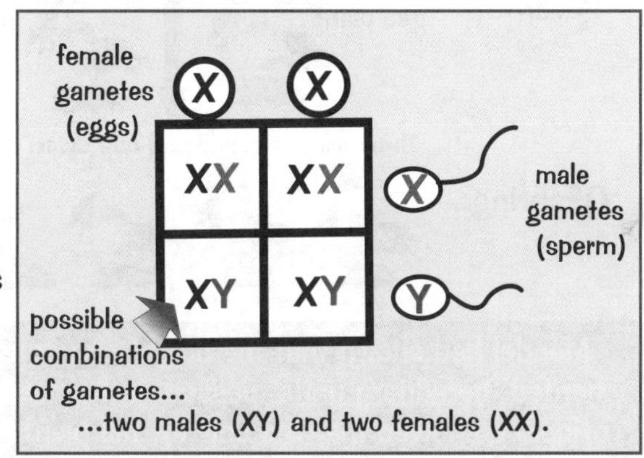

The other type of genetic diagram looks a bit more complicated, but it shows exactly the same thing.

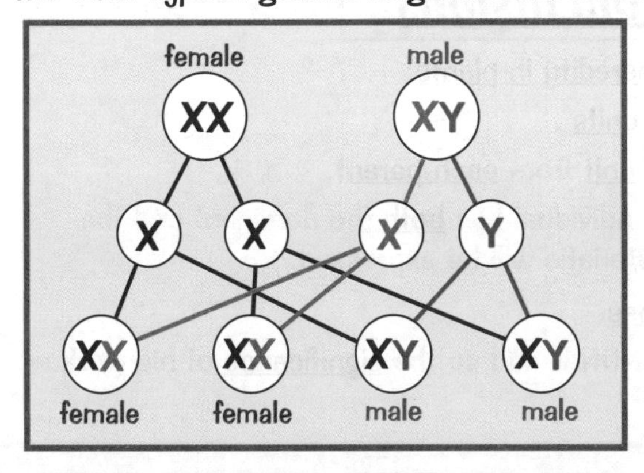

At the top are the parents.

The middle circles show the possible gametes that are formed. One gamete from the female combines with one gamete from the male (during fertilisation).

The criss-cross lines show all the possible ways the X and Y chromosomes could combine. The possible combinations of the offspring are shown in the bottom circles.

Remember, only one of these possibilities would actually happen for any one offspring.

Have you got the Y-factor...

Most genetic diagrams you'll see in exams concentrate on a gene, instead of a chromosome.
But the principle's the same. Don't worry — there are loads of other examples on the following pages.

The Work of Mendel

Mendel Did Genetic Experiments with Pea Plants

Gregor Mendel was an Austrian monk who trained in mathematics and natural history at the University of Vienna. On his garden plot at the monastery, Mendel noted how characteristics in plants were passed on from one generation to the next.

The results of his research were published in 1866 and eventually became the foundation of modern genetics.

The diagrams show two crosses for height in pea plants that Mendel carried out...

First Cross

A tall pea plant and a dwarf pea plant are crossed

Parents: Tall pea plant — Dwarf pea plant

All tall pea plants

Offspring:

Second Cross

Two pea plants from the 1st set of offspring are crossed

Parents: Tall pea plant — Tall pea plant

Three tall pea plants and one dwarf pea plant

Offspring:

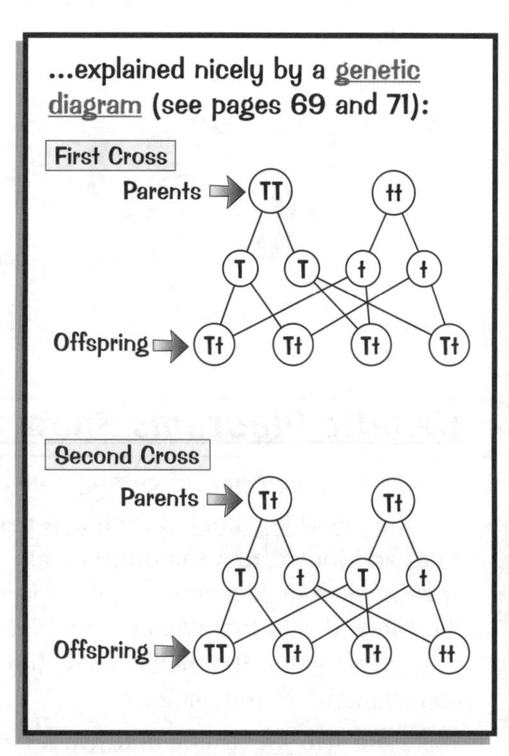

Mendel had shown that the height characteristic in pea plants was determined by separately inherited "hereditary units" passed on from each parent. The ratios of tall and dwarf plants in the offspring showed that the unit for tall plants, T, was dominant over the unit for dwarf plants, t.

Mendel Reached Three Important Conclusions

Mendel reached these three important conclusions about heredity in plants:

1) Characteristics in plants are determined by "hereditary units".
2) Hereditary units are passed on from both parents, one unit from each parent.
3) Hereditary units can be dominant or recessive — if an individual has both the dominant and the recessive unit for a characteristic, the dominant characteristic will be expressed.

We now know that the "hereditary units" are of course genes.

But in Mendel's time nobody knew anything about genes or DNA, and so the significance of his work was not to be realised until after his death.

Clearly, being a monk in the 1800s was a right laugh...

Well, there was no TV in those days, you see. Monks had to make their own entertainment. And in Mendel's case, that involved growing lots and lots of peas. He was a very clever lad, was Mendel, but unfortunately just a bit ahead of his time. Nobody had a clue what he was going on about.

Genetic Diagrams

In the exam they could ask about the inheritance of any kind of characteristic that's controlled by a single gene, because the principle's always the same. So here's a slightly bizarre example, to show you the basics.

Genetic Diagrams Show the Possible Genes of Offspring

1) Alleles are different versions of the same gene.
2) Most of the time you have two copies of each gene — one from each parent.
3) If they're different alleles, only one might be 'expressed' in the organism. The characteristic that appears is coded for by the dominant allele. The other one is recessive.
4) In genetic diagrams letters are used to represent genes. Dominant alleles are always shown with a capital letter, and recessive alleles with a small letter.

You Need to be Able to Interpret, Explain and Construct Them

Imagine you're cross-breeding hamsters, some with normal hair and a mild disposition and others with wild scratty hair and a leaning towards crazy acrobatics.

We'll use the letter 'B' (for boring) to represent the gene — always choose a letter whose capital looks different from the lower case one, so the examiners know exactly which one you're writing.

Let's say that the allele which causes the crazy nature is recessive, so we use a small "b" for it, whilst normal (boring) behaviour is due to a dominant allele, so we represent it with a capital "B".

1) For an organism to display a recessive characteristic, both its alleles must be recessive — so a crazy hamster must have the alleles 'bb'.
2) However, a normal hamster can have two possible combinations of alleles, BB or Bb, because the dominant allele overrules the recessive one.

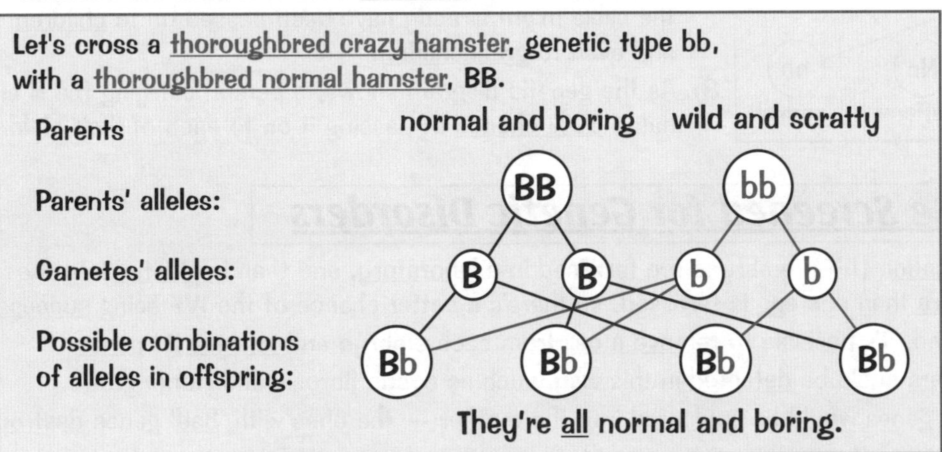

If two of these offspring now breed they will produce a new combination of kids.

Parents: normal and boring, normal and boring

Parents' alleles: Bb, Bb

Gametes' alleles: B, b, B, b

Possible combinations of alleles in offspring: BB, Bb, Bb, bb

normal, normal, normal, crazy!

It's not just hamsters that have the wild and scratty allele...

...my sister definitely has it too. Remember, "results" like this are only probabilities. It doesn't mean it'll actually happen. (Most likely, you'll end up trying to contain a mini-riot of nine lunatic baby hamsters.)

Genetic Disorders

Defective genes can cause serious problems — you need to know about two of them.

Cystic Fibrosis is Caused by a Recessive Allele

Cystic fibrosis is a genetic disorder of the cell membranes. It results in the body producing a lot of thick sticky mucus in the air passages and in the pancreas.

1) The allele which causes cystic fibrosis is a recessive allele, 'f', carried by about 1 person in 30.
2) Because it's recessive, people with only one copy of the allele won't have the disorder — they're known as carriers.
3) For a child to have the disorder, both parents must be either carriers or sufferers.
4) As the diagram shows there's a 1 in 4 chance of a child having the disease if both parents are carriers.

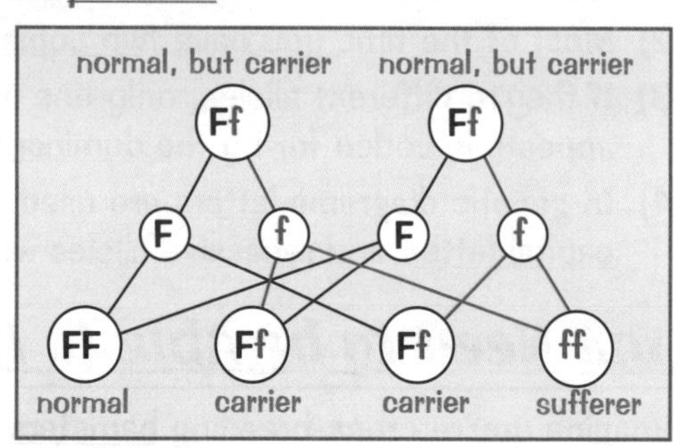

Huntington's is Caused by a Dominant Allele

Huntington's is a genetic disorder of the nervous system that's really horrible, resulting in shaking, erratic body movements and eventually severe mental deterioration.

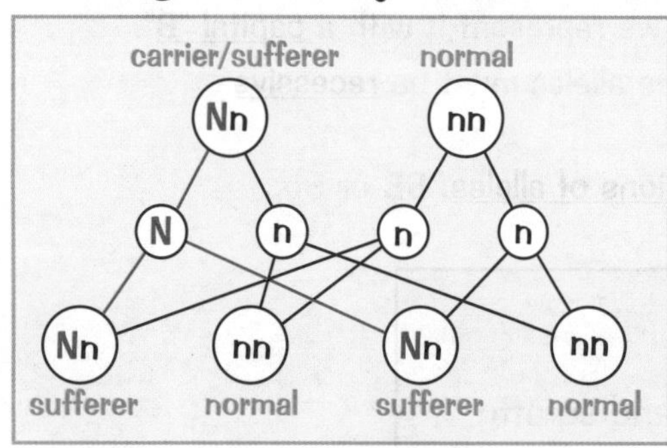

1) The disorder is caused by a dominant allele, 'N', and so can be inherited if just one parent carries the defective gene.
2) The "carrier" parent will of course be a sufferer too since the allele is dominant, but the symptoms don't start to appear until after the person is about 40. By this time the allele might already have been passed on to children and even to grandchildren.
3) As the genetic diagram shows, a person carrying the N allele has a 50% chance of passing it on to each of their children.

Embryos Can Be Screened for Genetic Disorders

1) During in vitro fertilisation (IVF), embryos are fertilised in a laboratory, and then implanted into the mother's womb. More than one egg is fertilised, so there's a better chance of the IVF being successful.
2) Before being implanted, it's possible to remove a cell from each embryo and analyse its genes.
3) Many genetic disorders could be detected in this way, such as cystic fibrosis and Huntington's.
4) Embryos with 'good' genes would be implanted into the mother — the ones with 'bad' genes destroyed.

There is a huge debate raging about embryonic screening. Here are some arguments for and against it.

Against Embryonic Screening

1) There may come a point where everyone wants to screen their embryos so they can pick the most 'desirable' one, e.g. they want a blue eyed, blonde haired, intelligent boy.
2) The rejected embryos are destroyed — they could have developed into humans.
3) It implies that people with genetic problems are 'undesirable' — this could increase prejudice.

For Embryonic Screening

1) It will help to stop people suffering.
2) There are laws to stop it going too far. At the moment parents cannot even select the sex of their baby (unless it's for health reasons).
3) During IVF, most of the embryos are destroyed anyway — screening just allows the selected one to be healthy.
4) Treating disorders costs the Government (and the taxpayers) a lot of money.

Embryonic screening — it's a tricky one...

There's a nice moral argument for you to consider on this page. In the exam you may be asked your opinion — make sure you can back it up with good reasons, and consider other points of view.

More Genetic Diagrams

You've got to be able to predict and explain the outcomes of crosses between individuals for each possible combination of dominant and recessive alleles of a gene. If you've got your head round all this genetics lark you should be able to draw a genetic diagram and work it out — but it'll make it easier if you've seen them all before. So here are some examples for you:

All the Offspring are Normal

Let's take another look at the crazy hamster example:

In this cross, a hamster with two dominant alleles (BB) is crossed with a hamster with two recessive alleles (bb). All the offspring are normal and boring.

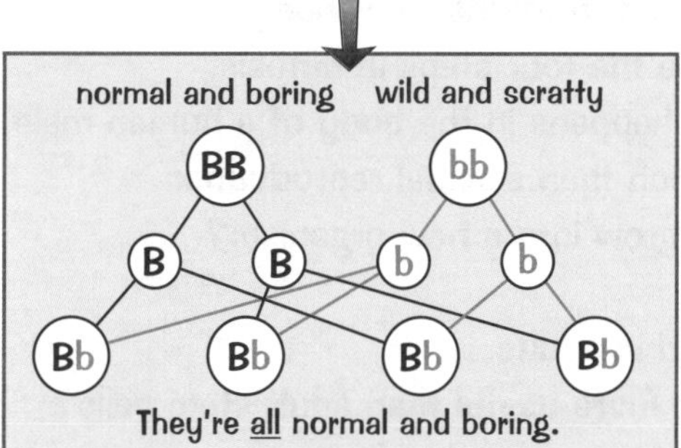

But, if you crossed a hamster with two dominant alleles (BB) with a hamster with a dominant and a recessive allele (Bb), you would also get all normal and boring offspring.

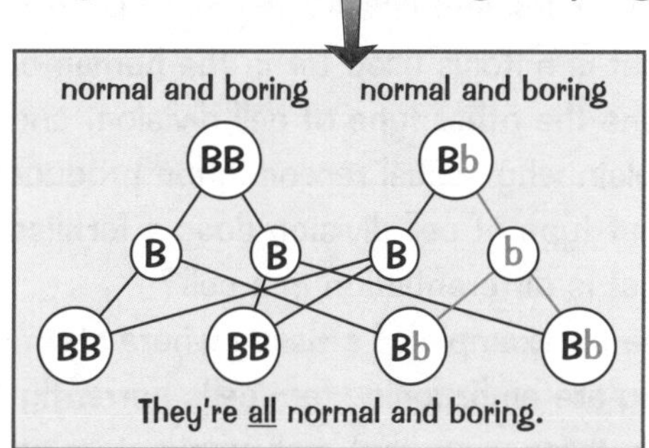

To find out which it was you'd have to breed the offspring together and see what kind of ratio you got that time — then you'd have a good idea. If it was 3:1, it's likely that you originally had BB and bb.

There's a 3:1 Ratio in the Offspring

Sickle cell anaemia is a genetic disorder characterised by funny-shaped red blood cells.

It's caused by a recessive allele 'a' (for anaemia). The normal allele is represented by an 'A'.

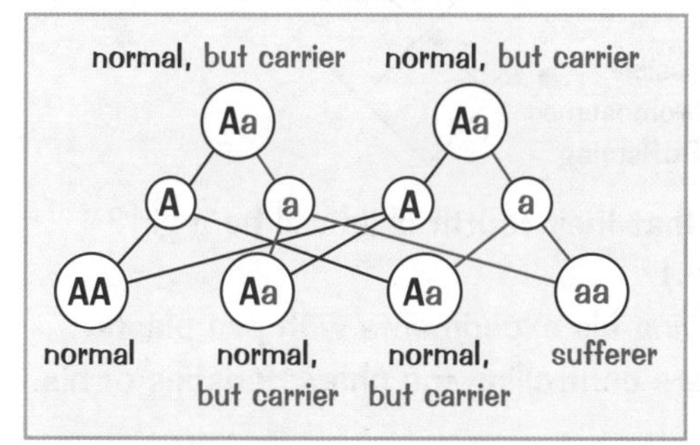

If two people who carry the sickle cell anaemia allele have children, the probability of each child suffering from the disorder is 1 in 4 — 25%.

The ratio you'd expect in the children is 3:1, non-sufferer:sufferer.

If you see this ratio in the offspring you know both parents must have the two different alleles.

Be careful with this one — it may be disguised as a 1:2:1 ratio (normal:carrier:sufferer), but it means the same thing.

There's a 1:1 Ratio in the Offspring

A cat with long hair was bred with another cat with short hair. The long hair is caused by a dominant allele 'H', and the short hair by a recessive allele 'h'.

They had 8 kittens — 4 with long hair and 4 with short hair.

This is a 1:1 ratio — it's what you'd expect when a parent with only one dominant allele (Hh) is crossed with a parent with two recessive alleles (hh).

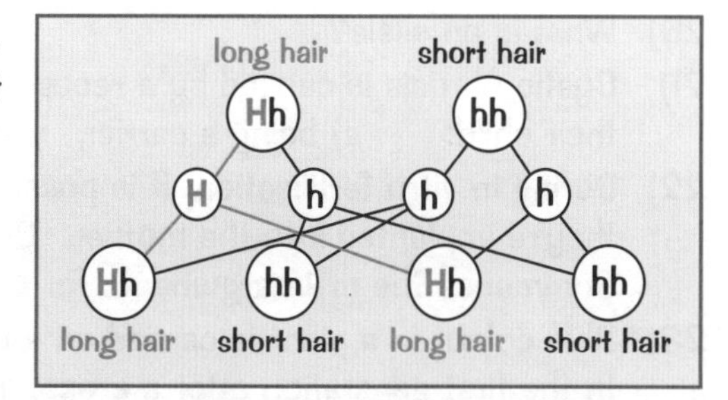

It's enough to make you go cross-eyed...

Remember that these are only probabilities, so you need loads of organisms in each generation to see a reliable ratio. That's why people tend to do genetic experiments with quick-breeding organisms like fruit flies. And of course it still won't be an exact ratio (just because of normal chance) — you might get 69 normal fruit flies and 31 crazy fruit flies out of 100. Not exact, but close enough to a 3:1 ratio.

Revision Summary for Biology 2(iii)

This section contains loads of new stuff all about genetics — it's even got a couple of moral dilemmas for you to ponder over. Your DNA contains all the instructions needed to make you. But your DNA can only account for so much, whether you can roll your tongue for example. You can't blame it for what colour you decide to dye your hair, or for leaving your smelly socks on the stairs though...

Use these questions to find out what you know about it all — and what you don't. Then look back and learn the bits you don't know. Then try the questions again, and again...

1) What is a gene?
2) Explain how DNA controls the activities of a cell.
3) Explain how DNA fingerprinting is used in forensic science.
4) Some people would like there to be a genetic database of everyone in the country. Discuss the advantages and disadvantages of such a database for use in forensic science.
5) What is mitosis used for in the human body? Describe the four steps in mitosis.
6) Name the other type of cell division, and say where it happens in the body of a human male.
7) Explain why sexual reproduction produces more variation than asexual reproduction.
8) What type of cell division does a fertilised egg use to grow into a new organism?
9) What is differentiation in a cell?
10) Give an example of a tissue where stem cells are found in adults.
11) Why are embryonic stem cells currently thought to be more useful than adult stem cells?
12) Give three ways that embryonic stem cells could be used to cure diseases.
13) Discuss the moral arguments for and against embryonic stem cell research in the UK.
14) Which chromosome in the human body causes male characteristics?
15) Copy and complete the diagrams below to show what happens to the X and Y chromosomes during reproduction.

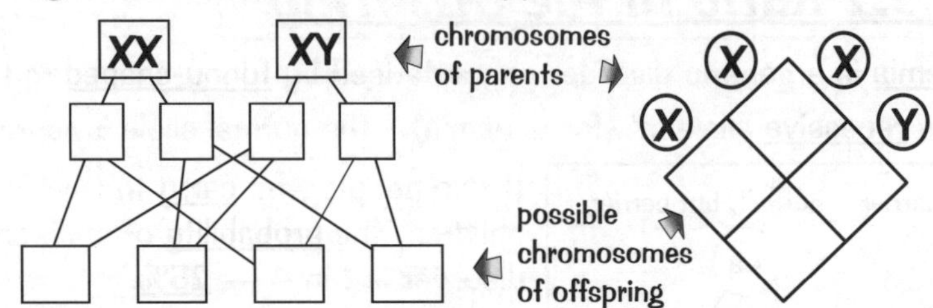

16)* A couple have three boys already. What is the probability that their fourth child will be a girl? (Hint: this may be a nasty trick question — don't be fooled.)
17) List three important conclusions that Mendel reached following his experiments with pea plants.
18) What were the "hereditary units" that Mendel concluded were controlling the characteristics of his pea plants?
19) The significance of Mendel's work was not realised until 1900, 16 years after Mendel died. Suggest why the importance of the work wasn't understood at the time.
20) What is an allele?
21) Cystic fibrosis is caused by a recessive allele. If both parents are carriers, what is the probability of their child: a) being a carrier, b) suffering from the disorder?
22) During in vitro fertilisation, it is possible to screen embryos for various genetic disorders before they're implanted into the mother. Only the "good" embryos would be chosen for implantation. Summarise the main arguments for and against embryonic screening.
23)* Blue colour in a plant is carried on a recessive allele, b. The dominant allele, B, gives white flowers. In the first generation after a cross, all the flowers are white. These are bred together and the result is a ratio of 54 white : 19 blue. What were the alleles of the flowers used in the first cross?

* Answers on page 96

Gas and Solute Exchange

The processes that keep organisms alive won't happen without the right raw materials. And the raw materials have to get to the right places. It's like making chicken soup. You need the chicken in your kitchen. It's no good if it's still at the supermarket.

Substances Move by Diffusion, Osmosis and Active Transport

1) Life processes need gases or other dissolved substances before they can happen.
2) For example, for photosynthesis to happen, carbon dioxide and water have to get into plant cells. And for respiration to take place, glucose and oxygen both have to get inside cells.
3) Waste substances also need to move out of the cells so that the organism can get rid of them.
4) These substances move to where they need to be by diffusion, osmosis and active transport.
5) Diffusion is where particles move from an area of high concentration to an area of low concentration. For example, different gases can simply diffuse through one another, like when a weird smell spreads out through a room. Alternatively, dissolved particles can diffuse in and out of cells through cell membranes — see page 41.
6) Osmosis is similar, but it only refers to water. The water moves across a partially permeable membrane (e.g. a cell membrane) from an area of high water concentration to an area of low water concentration — see page 42.
7) Diffusion and osmosis both involve stuff moving from an area where there's a high concentration of it, to an area where there's a lower concentration of it. Sometimes substances need to move in the other direction — which is where active transport comes in — see page 76.
8) In life processes, the gases and dissolved substances have to move through some sort of exchange surface. The exchange surface structures have to allow enough of the necessary substances to pass through.

Exchange surfaces are ADAPTED to maximise effectiveness.

The Structure of Leaves Lets Gases Diffuse In and Out of Cells

1) Carbon dioxide diffuses into the air spaces within the leaf, then it diffuses into the cells where photosynthesis happens. The leaf's structure is adapted so that this can happen easily.
2) The underneath of the leaf is an exchange surface. It's covered in biddy little holes called stomata which the carbon dioxide diffuses in through.
3) Water vapour and oxygen also diffuse out through the stomata. (Water vapour is actually lost from all over the leaf surface, but most of it is lost through the stomata.)
4) The size of the stomata are controlled by guard cells — see page 40. These close the stomata if the plant is losing water faster than it is being replaced by the roots. Without these guard cells the plant would soon wilt.
5) The flattened shape of the leaf increases the area of this exchange surface so that it's more effective.
6) The walls of the cells inside the leaf form another exchange surface. The air spaces inside the leaf increase the area of this surface so there's more chance for carbon dioxide to get into the cells.

The water vapour escapes by diffusion because there's a lot of it inside the leaf and less of it in the air outside. This diffusion is called transpiration and it goes quicker when the air around the leaf is kept dry — i.e. transpiration is quickest in hot, dry, windy conditions — and don't you forget it!

I say stomaaarta, you say stomaaayta...

The cells on the stem of a cactus photosynthesise and have stomata-like holes to let gases in. The cacti don't want to lose much water, so the holes only open at night when it's cooler. The cacti are adapted so that they can store the CO_2 that diffuses in at night until daylight when it's used for photosynthesis.

The Breathing System

You need to get oxygen from the air into your bloodstream so that it can get to your cells for respiration. You also need to get rid of carbon dioxide in your blood. This all happens inside the lungs. Breathing is how the air gets in and out of your lungs. Breathing's definitely a useful skill to have. You'll need to be able to do it to get through the exam.

The Lungs Are in the Thorax

1) The thorax is the top part of your 'body'.
2) It's separated from the lower part of the body by the diaphragm.
3) The lungs are like big pink sponges. and are protected by the ribcage.
4) The air that you breathe in goes through the trachea. This splits into two tubes called 'bronchi' (each one is 'a bronchus'), one going to each lung.
5) The bronchi split into progressively smaller tubes called bronchioles.
6) The bronchioles finally end at small bags called alveoli where the gas exchange takes place.

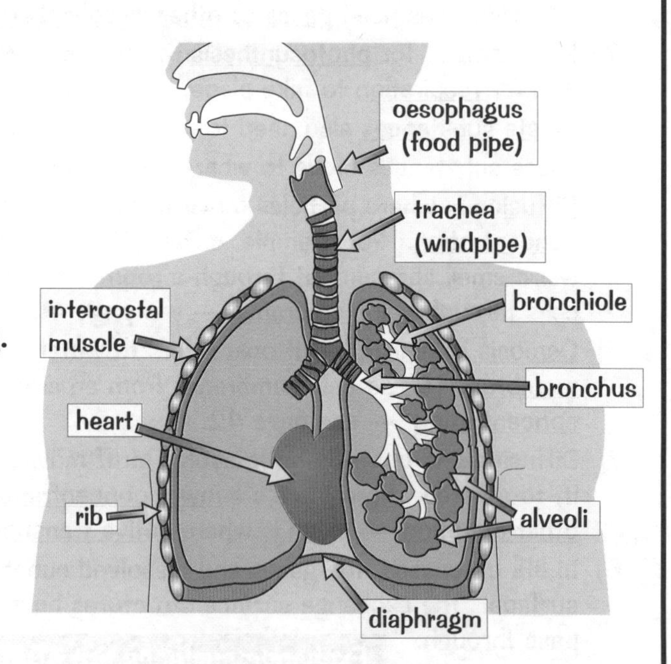

Remember — breathing in and out is DIFFERENT from respiration. See page 54.

Breathing In...

1) Intercostal muscles and diaphragm contract.
2) Thorax volume increases.
3) This decreases the pressure, drawing air in.

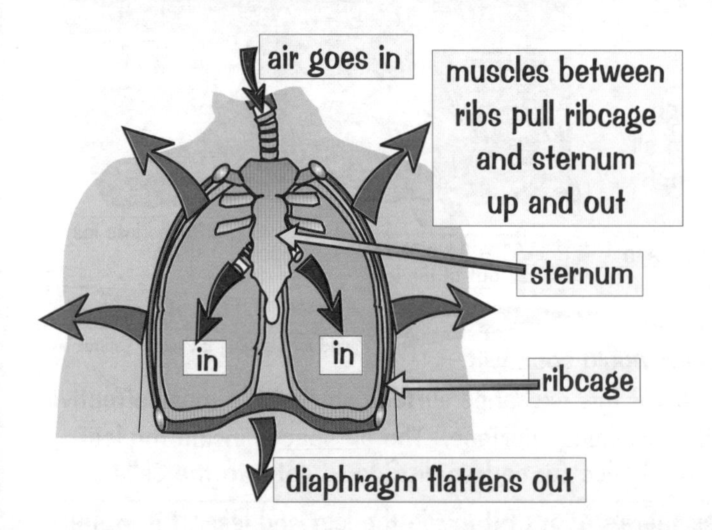

...and Breathing Out

1) Intercostal muscles and diaphragm relax.
2) Thorax volume decreases.
3) Air is forced out.

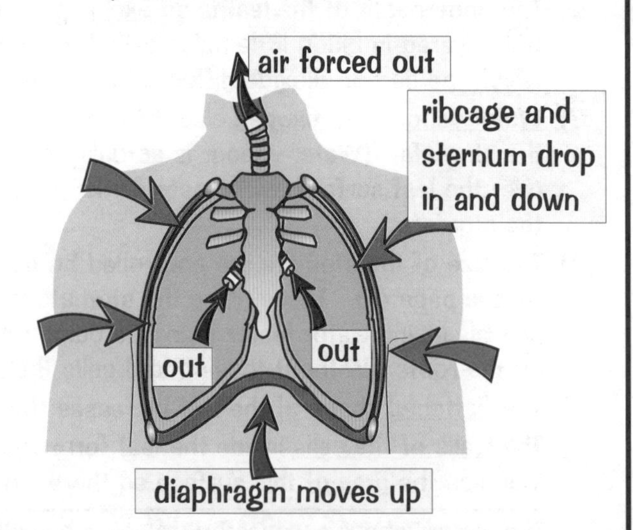

Stop huffing and puffing and just LEARN IT...

So when you breathe in, you don't have to suck the air in. You just make the space in your lungs bigger and the air rushes in to fill it. The small bags called alveoli at the ends of the air passages are the really interesting bit. It's through the alveoli that the oxygen gets into the blood supply to be carted off round the body. Also, the waste carbon dioxide gets out of the blood supply here so it can be breathed out. This is all explained on the next page, so once you've got this one learned, flip over and off you trot.

Diffusion Through Cell Membranes

This page is about how two different parts of the human body are adapted so that substances can diffuse through them most effectively. The first bit is about how gases in the lungs get into and out of the blood. The second is about how digested food gets from the gut to the blood.

Gas Exchange Happens in the Lungs

The job of the lungs is to transfer oxygen to the blood and to remove waste carbon dioxide from it. To do this the lungs contain millions of little air sacs called alveoli where gas exchange takes place.

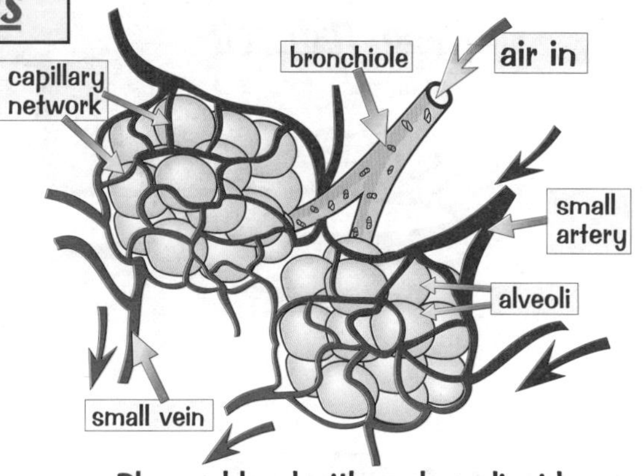

Blue = blood with carbon dioxide.
Red = blood with oxygen.

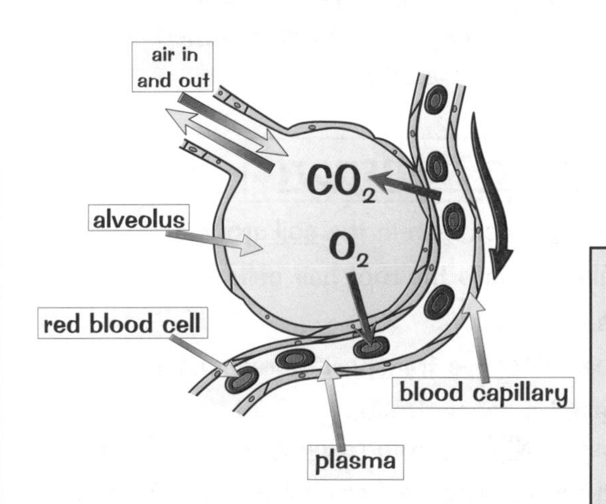

The alveoli are specialised to maximise the diffusion of oxygen and CO_2. They have:

- An enormous surface area (about $75m^2$ in humans).
- A moist lining for dissolving gases.
- Very thin walls.
- A copious blood supply.

The Villi Provide a Really Really Big Surface Area

The inside of the small intestine is covered in millions and millions of these tiny little projections called villi. They increase the surface area in a big way so that digested food is absorbed much more quickly into the blood.

Notice they have
- a single layer of surface cells
- a very good blood supply to assist quick absorption.

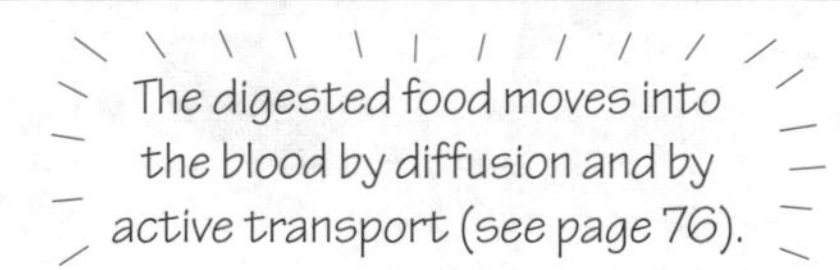
The digested food moves into the blood by diffusion and by active transport (see page 76).

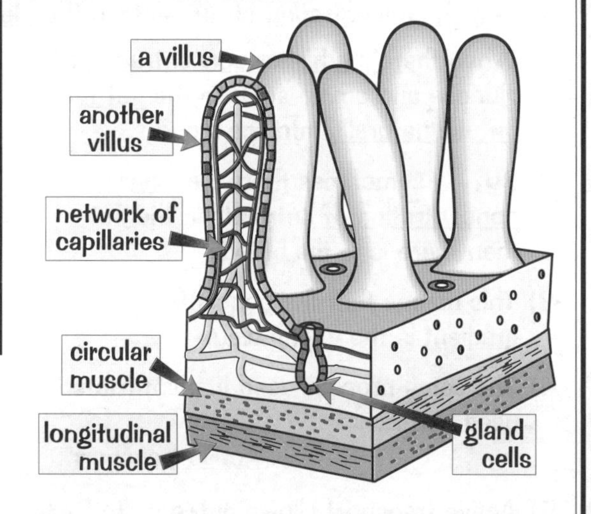

Al Veoli — the Italian gas man...

Living organisms are really well adapted for getting the substances they need to their cells. It makes sense — if they couldn't do this well, they'd die out. A large surface area is a key way that organisms' exchange surfaces are made more effective — molecules can only diffuse through a membrane when they're right next to it, and a large surface area means loads more molecules are close to the membrane. If you're asked how something's adapted for a job, think about if surface area is important.

Active Transport

Sometimes substances need to be absorbed against a concentration gradient, i.e. from a lower to a higher concentration. This process is lovingly referred to as ACTIVE TRANSPORT.

Root Hairs are Specialised for Absorbing Water and Minerals

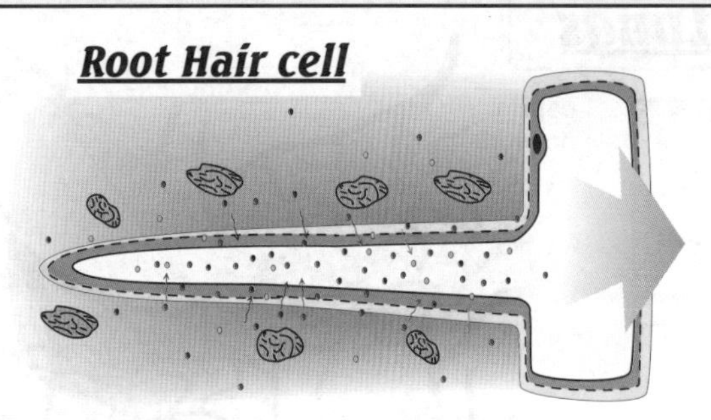

1) The cells on the surface of plant roots grow into long "hairs" which stick out into the soil.
2) This gives the plant a big surface area for absorbing water and mineral ions from the soil.
3) Most of the water and mineral ions that get into a plant are absorbed by the root hair cells.

Root Hairs Take in Minerals Using Active Transport

1) The concentration of minerals is usually higher in the root hair cell than in the soil around it.
2) So normal diffusion doesn't explain how minerals are taken up into the root hair cell.
3) They should go the other way if they followed the rules of diffusion.
4) The answer is that a conveniently mysterious process called "active transport" is responsible.
5) Active transport allows the plant to absorb minerals against a concentration gradient. This is essential for its growth. But active transport needs ENERGY from respiration to make it work.
6) Active transport also happens in humans, for example in taking glucose from the gut (see below), and from the kidney tubules.

We Need Active Transport to Stop Us Starving

Active transport is used in the gut when there is a low concentration of nutrients in the gut, but a high concentration of nutrients in the blood.

1) When there's a higher concentration of glucose and amino acids in the gut they diffuse naturally into the blood.
2) BUT — sometimes there's a lower concentration of nutrients in the gut than there is in the blood.
3) This means that the concentration gradient is the wrong way.
4) The same process used in plant roots is used here....

...."Active transport".

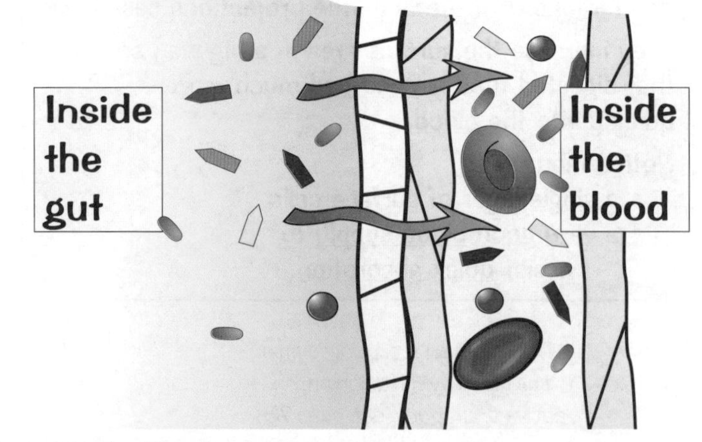

5) Active transport allows nutrients to be taken into the blood, despite the fact that the concentration gradient is the wrong way.

Active transport sucks...

An important difference between active transport and diffusion is that active transport uses energy. Imagine a pen of sheep in a field. If you open the pen, the sheep will happily diffuse from the area of high sheep concentration into the field, which has a low sheep concentration — you won't have to do a thing. To get them back in the pen though, you'll have to put in quite a bit of energy.

The Circulation System

The circulation system's main function is to get food and oxygen to every cell in the body.
As well as being a delivery service, it's also a waste collection service — it carries waste products like carbon dioxide and urea to where they can be removed from the body.

The DOUBLE Circulation System, Actually

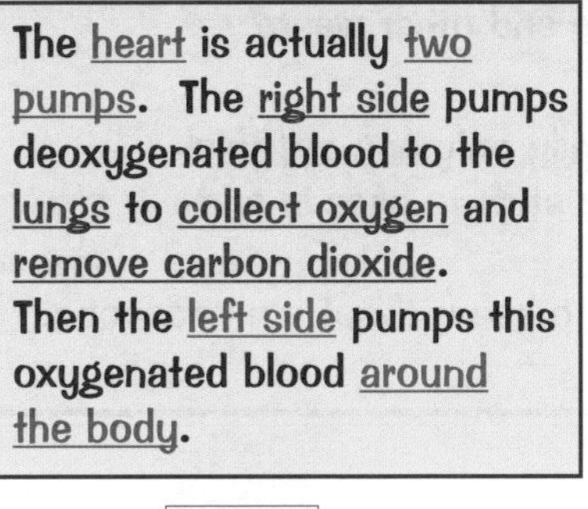

The heart is actually two pumps. The right side pumps deoxygenated blood to the lungs to collect oxygen and remove carbon dioxide. Then the left side pumps this oxygenated blood around the body.

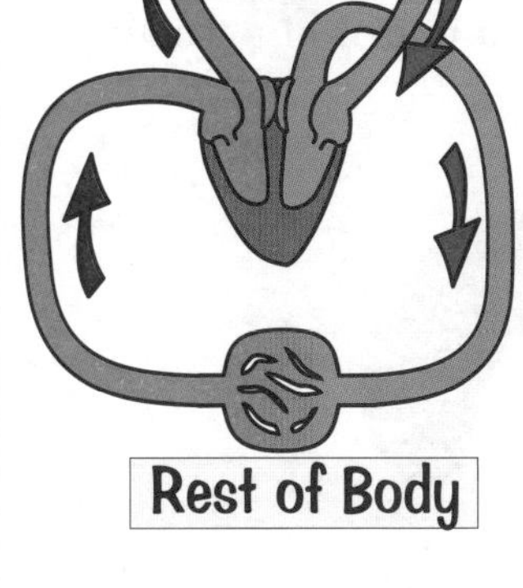

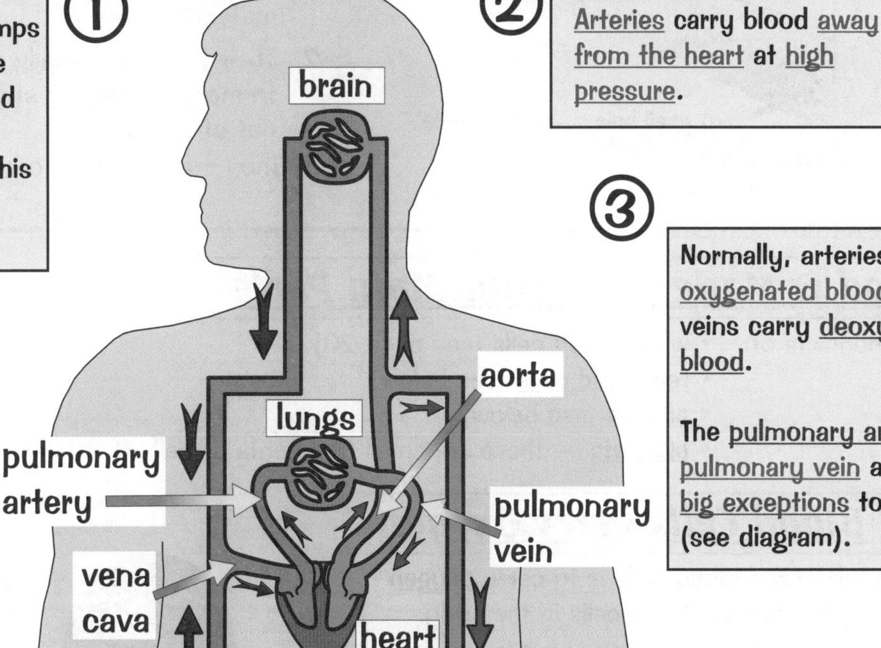

② Arteries carry blood away from the heart at high pressure.

③ Normally, arteries carry oxygenated blood and veins carry deoxygenated blood.

The pulmonary artery and pulmonary vein are the big exceptions to this rule (see diagram).

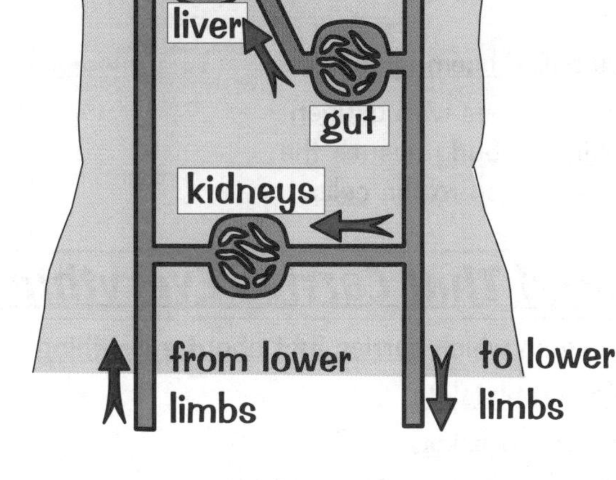

④ The arteries eventually split off into thousands of tiny capillaries which take blood to every cell in the body.

⑤ The veins then collect the "used" blood and carry it back to the heart at low pressure to be pumped round again.

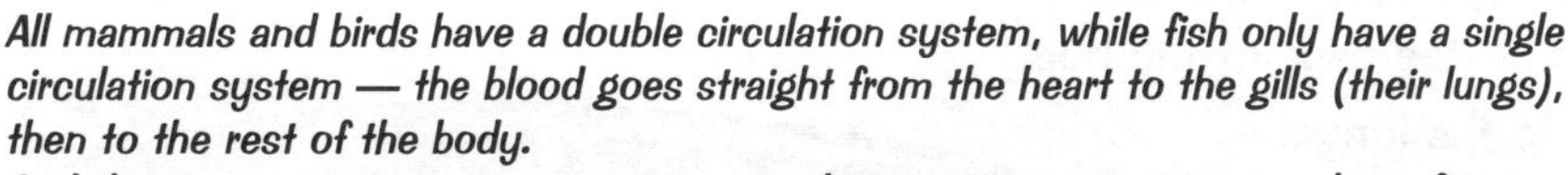

All mammals and birds have a double circulation system, while fish only have a single circulation system — the blood goes straight from the heart to the gills (their lungs), then to the rest of the body.
And there are some even more curious circulation systems, e.g. worms have five pairs of hearts, (which seems a bit silly to me as that's nine more that can be broken), and flat worms don't actually have any circulation system. It's all very odd. But then life's full of little mysteries isn't it!

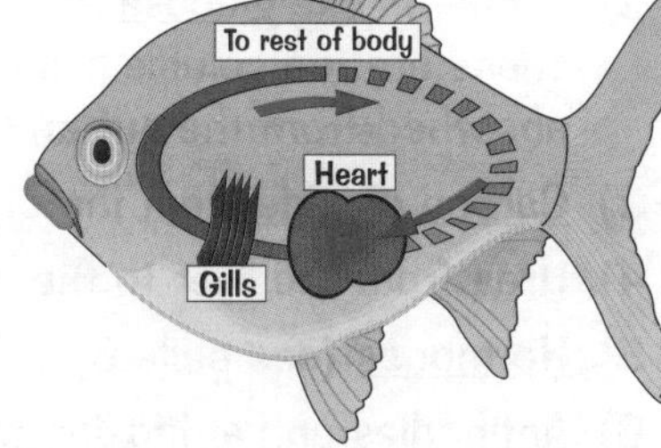

Blood vessels — a vampire's favourite type of ship...

The diagram above only shows the basic layout. There's actually zillions of blood vessels. If you laid all your arteries, capillaries and veins end to end, they'd go around the world about three times. These vessels vary from hose-pipe width arteries to capillaries that are a tenth of the thickness of a human hair.

Blood

This stuff's the nitty gritty about blood. Make sure you learn it all.

Capillaries Deliver Food and Oxygen to Each Cell

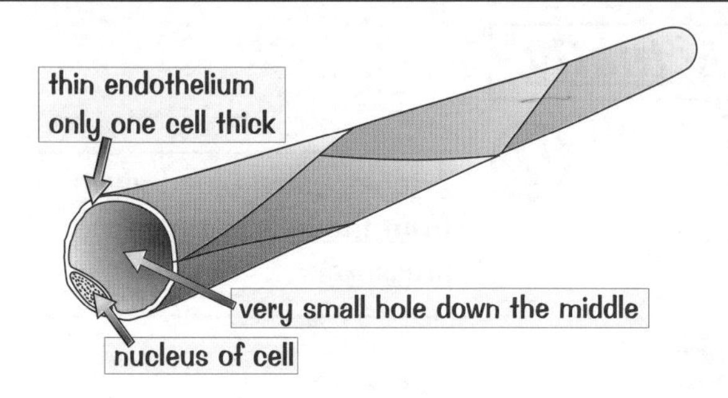

1) Capillaries use diffusion to deliver food and oxygen direct to body tissues and take carbon dioxide and other waste materials away.
2) Their walls are usually only one cell thick to make it easy for stuff to pass in and out of them.
3) They are too small to see without a microscope.

Blood is Made Up of Four Main Parts

Blood consists of:
- white blood cells (see page 20)
- red blood cells (see below)
- plasma (see below)
- platelets — these are small fragments of cells that help blood to clot at a wound.

Red Blood Cells Carry Oxygen

1) The job of red blood cells is to carry oxygen from the lungs to all the cells in the body.
2) They have a doughnut shape to give a large surface area for absorbing oxygen.
3) They don't have a nucleus — this allows more room to carry oxygen.
4) They contain a substance called haemoglobin.
5) In the lungs, haemoglobin combines with oxygen to become oxyhaemoglobin. In body tissues the reverse happens to release oxygen to the cells.

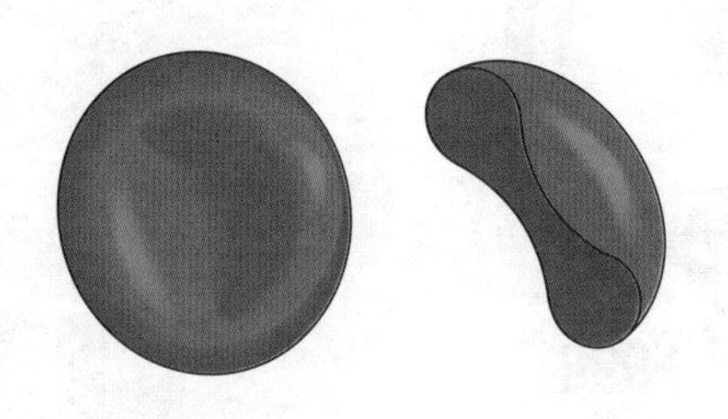

The more red blood cells you've got, the more oxygen can get to your cells. At high altitudes there's less oxygen in the air — so people who live there produce more red blood cells to compensate.

Plasma is the Liquid That Carries Everything in Blood

This is a pale straw-coloured liquid which carries just about everything:
1) Red and white blood cells and platelets.
2) Nutrients like glucose and amino acids. These are the soluble products of digestion which are absorbed from the gut and taken to the cells of the body.
3) Carbon dioxide from the organs to the lungs.
4) Urea from the liver to the kidneys.
5) Hormones (see page 8).
6) Antibodies and antitoxins produced by the white blood cells.

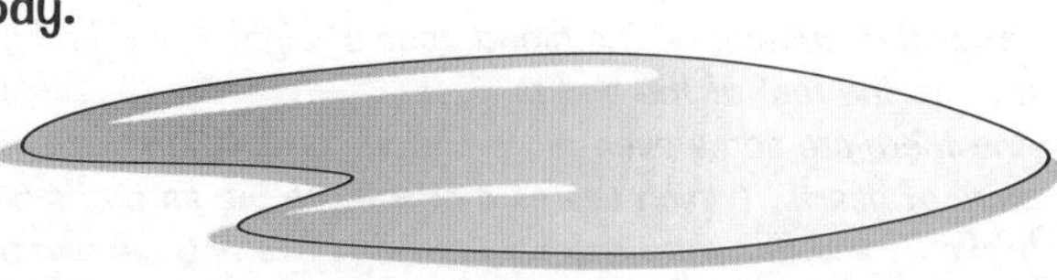

Advice for a vampire — drink your soup before it clots...

Red blood cells are perfectly designed for absorbing plenty of oxygen and squeezing through capillaries. There's a condition called sickle-cell anaemia in which the red blood cells are crescent-moon shapes. This causes problems because less oxygen is carried and the cells don't flow well through the capillaries.

Exercise

When you exercise, your body quickly adapts so that your muscles get more oxygen and glucose to supply energy. If your body can't get enough oxygen or glucose to them, it has some back-up plans ready.

Exercise Increases the Heart Rate

1) Muscles are made of muscle cells. These use oxygen to release energy from glucose (this process is called respiration), which is used to contract the muscles.

See page 54.

2) An increase in muscle activity requires more glucose and oxygen to be supplied to the muscle cells. Extra carbon dioxide needs to be removed from the muscle cells. For this to happen the blood has to flow at a faster rate.
3) This is why physical activity:
   - increases your breathing rate and makes you breathe more deeply to meet the demand for extra oxygen.
   - increases the speed at which the heart pumps.
   - dilates (makes wider) the arteries which supply blood to the muscles.

Glycogen is Used During Exercise

1) Some glucose from food is stored as glycogen.
2) Glycogen's mainly stored in the liver, but each muscle also has its own store.
3) During vigorous exercise, muscles use glucose rapidly, and have to draw on their glycogen stores to provide more energy. If the exercise goes on for a while the glycogen stores get used up.
4) When the glycogen stores run low, the muscles don't get the energy they need to keep contracting, and they get tired.

Anaerobic Respiration is Used if There's Not Enough Oxygen

1) When you do vigorous exercise and your body can't supply enough oxygen to your muscles, they start doing anaerobic respiration instead of aerobic respiration.
2) "Anaerobic" just means "without oxygen". It's the incomplete breakdown of glucose, which produces lactic acid.

   glucose → energy + lactic acid

3) This is NOT the best way to convert glucose into energy because lactic acid builds up in the muscles, which gets painful. This also causes the muscles to get tired.
4) Another downside is that anaerobic respiration does not release nearly as much energy as aerobic respiration — but it's useful in emergencies.
5) The advantage is that at least you can keep on using your muscles for a while longer.

Anaerobic Respiration Leads to an Oxygen Debt

1) After resorting to anaerobic respiration, when you stop exercising you'll have an "oxygen debt".
2) In other words you have to "repay" the oxygen that you didn't get to your muscles in time, because your lungs, heart and blood couldn't keep up with the demand earlier on.
3) This means you have to keep breathing hard for a while after you stop, to get oxygen into your muscles to oxidise the painful lactic acid to harmless CO_2 and water.
4) While high levels of CO_2 and lactic acid are detected in the blood (by the brain), the pulse and breathing rate stay high to try and rectify the situation.

Too much keep-fit makes your head spin round — like in the Exorcist...

Phew... I bet you're exhausted after reading this page. Yeast also respire anaerobically, but they produce ethanol (and carbon dioxide) — see page 84. It's a good job that humans produce lactic acid instead — or after a bit of vigorous exercise we'd all be staggering around drunk.

Kidneys

The kidneys are really important organs. They get rid of toxic waste like urea as well as adjusting the amount of dissolved ions and water in the blood. The kidneys were introduced on page 59, but here's the rest of the stuff you need to know.

Nephrons Are the Filtration Units in the Kidneys

1) Ultrafiltration:

1) A high pressure is built up which squeezes water, urea, ions and sugar out of the blood and into the Bowman's capsule.
2) The membranes between the blood vessels and the Bowman's capsule act like filters, so big molecules like proteins and blood cells are not squeezed out. They stay in the blood.

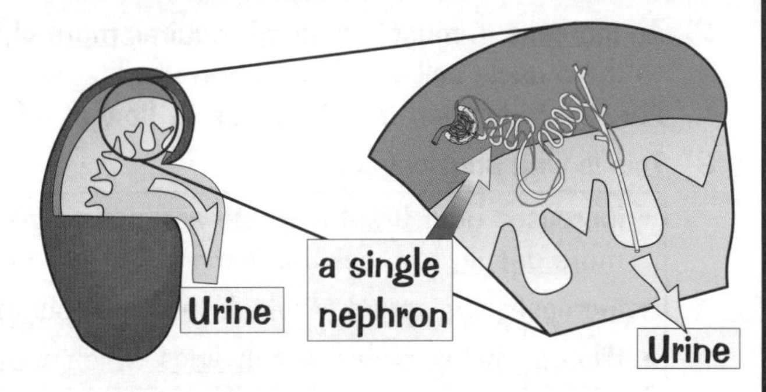

Enlarged View of a Single Nephron

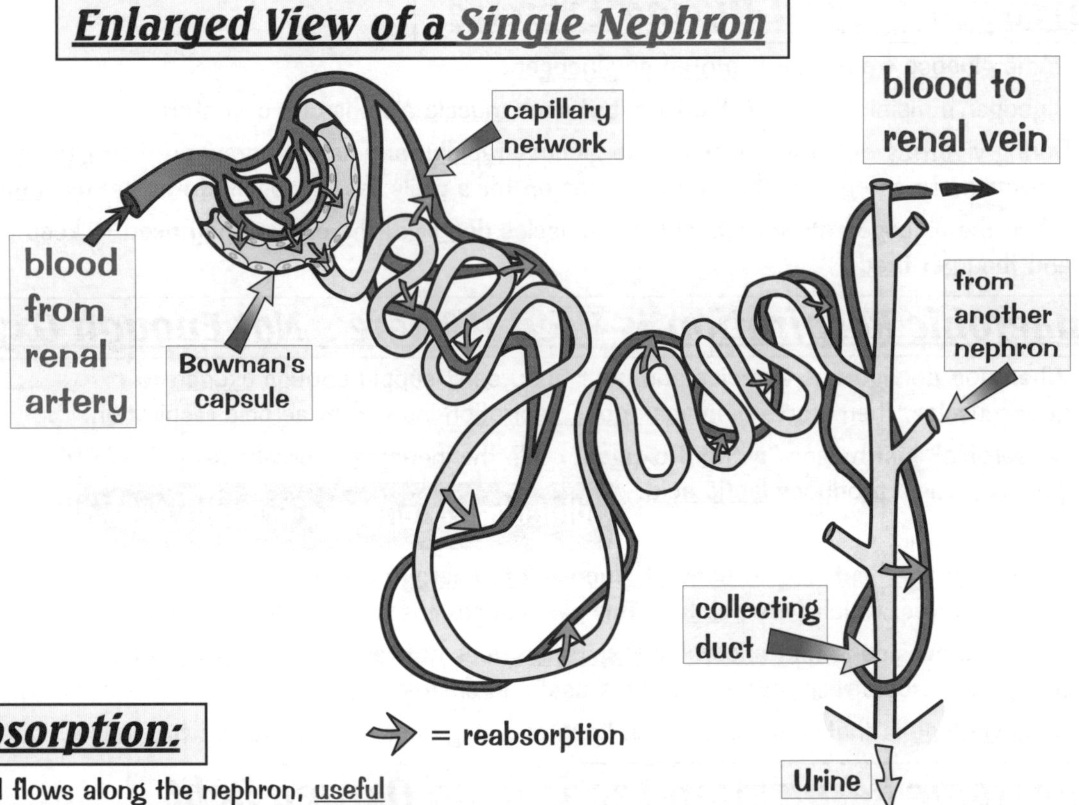

2) Reabsorption:

As the liquid flows along the nephron, useful substances are reabsorbed back into the blood:

1) All the sugar is reabsorbed. This involves the process of active transport against the concentration gradient.
2) Sufficient ions are reabsorbed. Excess ions are not. Active transport is needed.
3) Sufficient water is reabsorbed.

3) Release of wastes:

The remaining substances (including urea) continue out of the nephron, into the ureter and down to the bladder as urine.

Don't try to kid-me that you know it all — learn it properly...

The kidneys are pretty complicated organs as you can see. Luckily you don't have to learn all the ins and outs of the diagram — but you do have to make sure you know exactly what happens in each of the three stages. Learn what's filtered, what's reabsorbed and what's released as urine.

Kidney Failure

If someone's kidneys stop working, there are basically two treatments — regular dialysis or a transplant.

The Kidneys Remove Waste Substances from the Blood

1) If the kidneys don't work properly, waste substances build up in the blood and you lose your ability to control the levels of ions and water in your body. Eventually, this results in death.
2) People with kidney failure can be kept alive by having dialysis treatment — where machines do the job of the kidneys. Or they can have a kidney transplant.

The kidneys are incredibly important — if they don't work as they should, you can get problems in the heart, bones, nervous system, stomach, mouth, etc.

Dialysis Machines Filter the Blood

1) Dialysis has to be done regularly to keep the concentrations of dissolved substances in the blood at normal levels, and to remove waste substances.
2) In a dialysis machine the person's blood flows alongside a selectively permeable barrier, surrounded by dialysis fluid. It's permeable to things like ions and waste substances, but not big molecules like proteins (just like the membranes in the kidney).
3) The dialysis fluid has the same concentration of dissolved ions and glucose as healthy blood.
4) This means that useful dissolved ions and glucose won't be lost from the blood during dialysis.
5) Only waste substances (such as urea) and excess ions and water diffuse across the barrier.
6) Many patients with kidney failure have to have a dialysis session three times a week. Each session takes 3-4 hours — not much fun.

dialysis fluid out
selectively permeable membrane
dialysis fluid in
Waste products diffuse out into dialysis fluid
from person
back to person

Transplanted Organs can be Rejected by the Body

At the moment, the only cure for kidney disease is to have a kidney transplant. Healthy kidneys are usually transplanted from people who have died suddenly, say in a car accident, and who are on the organ donor register or carry a donor card (provided their relatives give the go-ahead). But kidneys can also be transplanted from people who are still alive — as we all have two of them.

The donor kidney can be rejected by the patient's immune system — treated like a foreign body and attacked by antibodies. To help prevent this happening, precautions are taken:

1) A donor with a tissue type that closely matches the patient is chosen. Tissue type is based on the antigens (see page 20) that are on the surface of most cells.
2) The patient's bone marrow is zapped with radiation to stop white blood cells being produced — so they won't attack the transplanted kidney. They also have to take drugs that suppress the immune system.
3) Unfortunately, this means that the patient can't fight any disease that comes along, so they have to be kept in totally sterile conditions for some time after the operation.

Dialysis or transplant? Both have their downsides...

Kidney dialysis machines are expensive things for the NHS to run — and dialysis is not a pleasant experience. Transplants can put an end to the hours spent on dialysis, but there are long waiting lists for kidneys. Even if one with a matching tissue type is found, there's the possibility that it'll be rejected. And taking drugs that suppress the immune system means the person is vulnerable to other illnesses.

Revision Summary for Biology 3(i)

It's no good just reading the section through and hoping you've got it all — it'll only stick if you've learned it properly. These questions are designed to really test whether you know all your stuff — ignore them at your peril. OK, rant over — I'll leave it to you...

1) What's the name for the process that's happening when water moves across a partially permeable membrane to equalise the concentrations on either side?
2) Explain how leaves are adapted to maximise the amount of carbon dioxide that gets to their cells.
3) Why do the leaves care if carbon dioxide gets to their cells or not?
4) What are the pores in the leaves called?
5) Name the main substances that diffuse out of leaves.
6) What conditions does transpiration happen most quickly in?
7) Cacti, which grow in the desert, have spikes instead of flat leaves. Why is this?
8) Name the chest cavity that's above the diaphragm.
9) Describe the gas exchange that happens between the alveoli and the blood.
10) Give four ways that the alveoli's structure is ideal for gas exchange.
11) How does the structure of a villus make it good at its job?
12) Give the two main differences between active transport and diffusion.
13) Why can't most mineral ions get into roots by diffusion?
14) Draw a diagram of a root hair cell. Why is it this shape?
15) Does glucose only get into the blood from the gut by active transport?
16) Explain why our circulation system is called a *double* circulation system.
17) Describe the pressure and oxygen content of the blood in veins and arteries. What are the big words for saying if the blood has oxygen in or not?
18) Sketch a red blood cell. Why is it this shape?
19) What's the substance in red blood cells called? What is it called when it combines with oxygen?
20)* Some companies sell special tents for athletes to sleep in. These tents have a lower oxygen concentration than the air at sea level has.
    a) Explain why an athlete might buy one of these tents.
    b) How could an athlete achieve the same effect without buying one of these tents?
21) Why does your heart beat faster when you do exercise?
22)* The table below shows the oxygen consumption of an athlete as her heart beats at different rates.

Heart rate (BPM)	85	105	118	125	143	148	152	163
Oxygen consumption (ml/kg/min)	13	26	30	33	40	47	53	56

    a) Draw a scattergraph of the data and a line of best fit.
    b) What does this graph show?
    c) Explain why this relationship exists.
23) What is "anaerobic respiration"? Give the word equation for what happens in our bodies.
24) Give two reasons why anaerobic respiration isn't the best way to release energy.
25) Explain how you repay an oxygen debt.
26) Name three things that are reabsorbed by kidneys.
27) Explain why sugar doesn't simply diffuse back into the blood from the nephrons.
28) How does kidney dialysis work?.
29) What are the advantages and disadvantages of a kidney transplant over dialysis?
30) Why do transplant patients have their immune systems suppressed?

* Answers on page 96

Food and Drink from Microorganisms

Microorganisms, such as bacteria, cause changes in food — often the changes are bad, but sometimes they're useful and mean we can have foods that we wouldn't have otherwise.

The Theory of Biogenesis Has Been Developed Over the Years

1) People used to think that life could spontaneously generate (just appear) from non-living material.
2) But then evidence showed that this couldn't be the case. The evidence supported the theory that living things are created from other living organisms — this is the theory of biogenesis.
3) Here's how the accepted theory was changed to fit the available evidence:

Before 1765 it was believed that substances in food were changed into microbes, which caused the food to go off.

A scientist called Lazzaro Spallanzani boiled two sets of broth to kill the microbes, then sealed one flask and left the other open. Only the open one went off (although the broth in the sealed flask did go off when it was left open later).

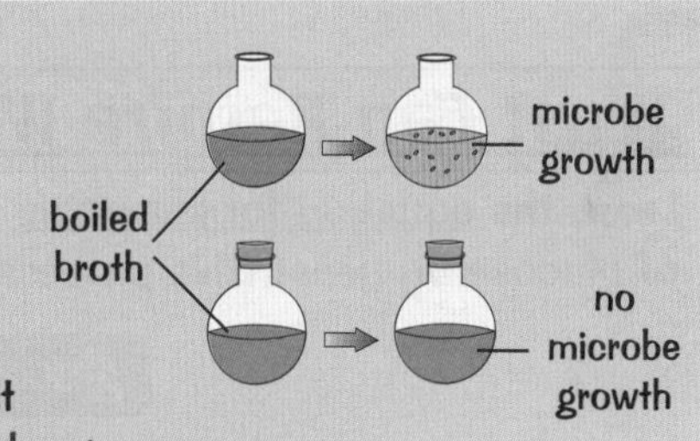

This showed that microbes got into the food from the air, but opponents just thought that it meant air from outside the flask was necessary to start the change.

The theory that 'fresh' air caused substances in food to change into microbes was disproved by Theodor Schwann in 1837. He showed that meat would not go off in air, provided the air was heated first to kill microorganisms.

A more conclusive experiment was carried out by the famous scientist Louis Pasteur in 1859.

He heated broth in two flasks, both of which were left open to the air. However, one of the flasks had a curved neck so that bacteria in the air would settle in the loop, and not get through to the broth.

The broth in the flask with the curved neck stayed fresh, proving that it was the microbes and not the air causing it to go off.

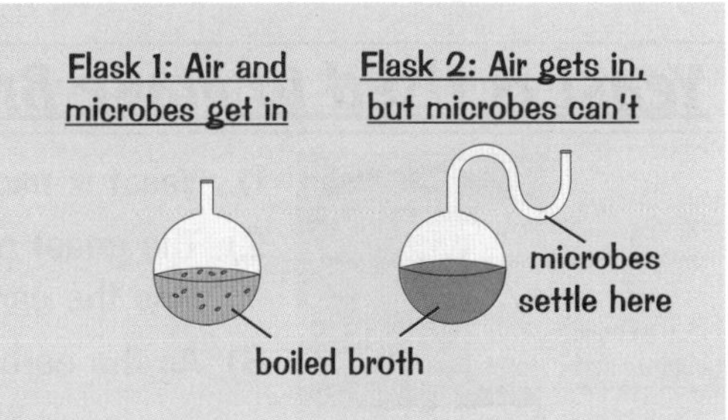

Most Cheese is Made Using Bacteria...

Yummy. Here's what happens:

1) A culture of bacteria is added to milk.
2) The bacteria produce solid curds in the milk.
3) The curds are separated from the liquid whey.
4) More bacteria are sometimes added to the curds, and the whole lot is left to ripen for a while.
5) Moulds are added to give blue cheese (e.g. Stilton) its colour and taste.

Yoghurt is Made Using Bacteria Too

Bacteria are used to clot milk during the manufacture of yoghurt.

1) The milk is often heat treated first to kill off any bacteria that may be in it, then cooled.
2) A starter culture of bacteria is then added. The bacteria ferment the lactose sugar (present in the milk) to lactic acid.
3) The acid causes the milk to clot and solidify into yoghurt.
4) Sterilised flavours (e.g. fruit) are sometimes then added.

So bacteria aren't always the bad guys...

It seems weird. Microorganisms in food can make you ill — that's why you should wash your hands before touching food and the reason you have to make sure meat is cooked thoroughly. Yet some foods are made with microorganisms, and as you'll see on page 85, some food is microorganisms.

Using Yeast

There's nothing newfangled about yeast. It's been used for donkey's years to make bread and alcohol.

Yeast is a Single-Celled Fungus

Yeast is a microorganism. A yeast cell has a nucleus, cytoplasm, a vacuole, and a cell membrane surrounded by a cell wall.

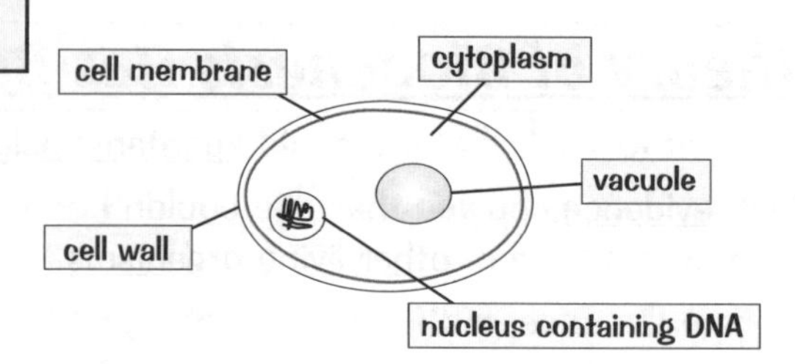

Yeast Can Respire With or Without Oxygen

Learn the equation for anaerobic respiration (i.e. without oxygen) of glucose by yeast (this process is called fermentation):

glucose → ethanol + carbon dioxide + energy

Yeast can also respire aerobically (i.e. with oxygen). This produces much more energy, and is needed to grow and reproduce:

glucose + oxygen → carbon dioxide + water + energy

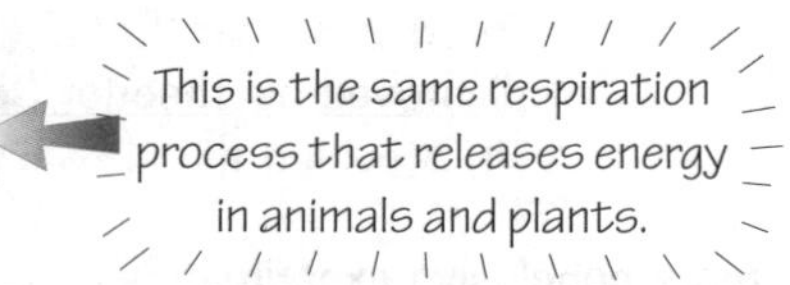

Yeast is Used to Make Bread

1) Yeast is used in dough to produce nice, light bread.
2) The yeast converts sugars to carbon dioxide and some ethanol. It is the carbon dioxide that makes the bread rise.
3) As the carbon dioxide expands, it gets trapped in the dough, making it lighter.

Yeast is Used to Make Alcoholic Drinks

Here's how beer is brewed:

1) Beer is made from grain — usually barley.
2) The barley grains are allowed to germinate for a few days, during which the starch in the grains is broken down into sugar by enzymes. Then the grains are dried in a kiln. This process is called malting.
3) The malted grain is mashed up and water is added to produce a sugary solution with lots of bits in it. This is then sieved to remove the bits.
4) Hops are added to the mixture to give the beer its bitter flavour.
5) The sugary solution is then fermented by yeast, turning the sugar into alcohol.

Germination is when a seed starts to grow into a new plant.

In wine-making, the yeast use the natural sugars in the grape juice as their energy source.

Invite yeast to parties — they're fun guys...

As some fruits get really ripe, the yeast start to ferment the natural sugars and make ethanol. If animals come along and guzzle down lots of the ripened fruit then they'll start to get a bit tipsy — no, seriously, it's true. I'd definitely want to keep out of the way of an elephant that was a bit unsteady on its feet.

Microorganisms in Industry

Shedloads of microorganisms are grown in huge vats called fermenters to make things like antibiotics, fuels and proteins. It's really important to control the conditions in fermenters so that just the stuff you want grows as fast as possible.

Microorganisms Are Grown in Fermenters on a Large Scale

A fermenter is a big container full of liquid culture medium which microorganisms can grow and reproduce in. The fermenter needs to give the microorganisms the conditions they need to grow and produce their useful product. The diagram shows a typical fermenter.

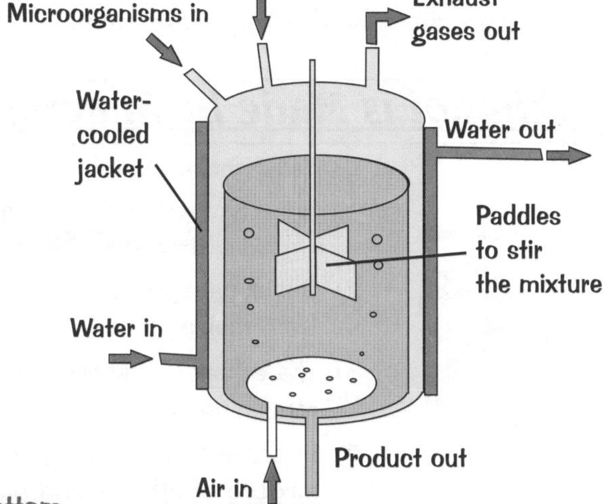

1) Food is provided in the liquid culture medium. More can be pumped in if needed.
2) Air is piped in to supply oxygen to the microorganisms.
3) The microorganisms need to be kept at the right temperature. The microorganisms produce heat by respiration, so the fermenters must be cooled. This is usually done with water in a water-cooled jacket. The temperature is monitored by instruments.
4) The right pH is needed for the microorganisms to thrive. Instruments will monitor this.
5) Sterile conditions are needed to prevent contamination from other microorganisms.
6) The microorganisms need to be kept from sinking to the bottom. A motorised stirrer keeps them moving around and maintains an even temperature.

Mycoprotein — Food from Fermenters

1) Mycoprotein means protein from fungi. It's a type of single-celled protein.
2) Mycoprotein is used to make meat substitutes for vegetarian meals — Quorn, for example.
3) A fungus called Fusarium is the main source of mycoprotein.
4) The fungus is grown in fermenters, using glucose syrup as food. The glucose syrup is obtained by digesting maize starch with enzymes.
5) The fungus respires aerobically, so oxygen is supplied, together with nitrogen (as ammonia) and other minerals.
6) It's important to prevent other microorganisms growing in the fermenter. So the fermenter is initially sterilised using steam. The incoming nutrients are heat sterilised and the air supply is filtered.

Penicillin is Made by Growing Mould in Fermenters

1) Penicillin is an antibiotic made by growing the mould *Penicillium chrysogenum* in a fermenter.
2) The mould is grown in a liquid culture medium containing sugar and other nutrients (for example, a source of nitrogen).
3) The sugar is used up as the mould grows.
4) The mould only starts to make penicillin after using up most of the nutrients for growth.

Alexander Fleming discovered Penicillin accidentally in 1928. A culture of bacteria became contaminated with a mould. This mould wiped out areas of bacteria. No one took much notice of Fleming's findings until the Second World War, when the huge number of injuries made it important to find something that would heal infected wounds.

Culture Medium — sounds very BBC Four to me...

Food made from microorganisms mightn't sound very appetising, but there are definitely advantages to it. In some developing countries it's difficult to find enough protein. Meat is a big source of protein, but animals need lots of space to graze, plenty of nice grass, etc. Single-celled protein grown in a fermenter is an efficient way of producing protein to feed people. The microorganisms grow very quickly, and don't need much space. And they can even feed on waste material that would be no good for feeding animals.

Fuels from Microorganisms

Food and antibiotics aren't the only things microorganisms can be used for — the stuff they produce can also be used as fuel. And with the world's oil and gas supplies running low, other fuel sources, such as this, are going to become really important.

Fuels Can Be Made by Fermentation

1) Fuels can be made by fermentation of natural products — luckily enough, waste products can often be used.
2) Fermentation is when bacteria or yeast break sugars down by anaerobic respiration.

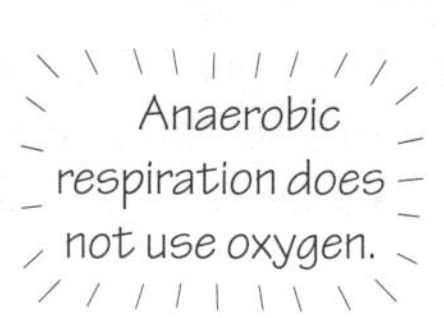

Ethanol is Made by Anaerobic Fermentation of Sugar

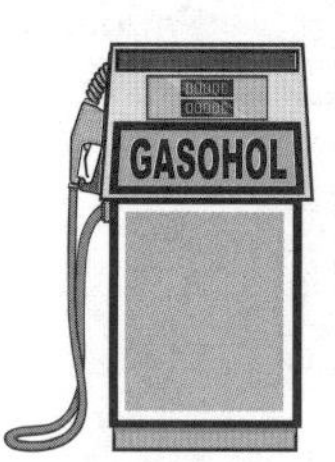

1) Yeast make ethanol when they break down glucose by anaerobic respiration.

Glucose → Ethanol + Carbon dioxide + Energy

This is the same as the reaction used in wine-making.

2) Sugar cane juices can be used, or glucose can be derived from maize starch by the action of carbohydrase (an enzyme).
3) The ethanol is distilled to separate it from the yeast and remaining glucose before it's used.
4) In some countries, e.g. Brazil, cars are adapted to run on a mixture of ethanol and petrol — this is known as 'gasohol'.

Biogas is Made by Anaerobic Fermentation of Waste Material

1) Biogas is usually about 70% methane (CH_4) and 30% carbon dioxide (CO_2).
2) Lots of different microorganisms are used to produce biogas. They ferment plant and animal waste, which contains carbohydrates. Sludge waste from, e.g. sewage works or sugar factories, is used to make biogas on a large scale.
3) It's made in a simple fermenter called a digester or generator (see the next page).
4) Biogas generators need to be kept at a constant temperature to keep the microorganisms respiring away.
5) There are two types of biogas generators — batch generators and continuous generators. These are explained on the next page.
6) Biogas can't be stored as a liquid (it needs too high a pressure), so it has to be used straight away — for heating, cooking, lighting, or to power a turbine to generate electricity.

Fuel Production Can Happen on a Large or Small Scale

1) Large-scale biogas generators are now being set up in a number of countries. Also, in some countries, small biogas generators are used to make enough gas for a village or a family to use in their cooking stoves and for heating and lighting.
2) Human waste, waste from keeping pigs, and food waste (e.g. kitchen scraps) can be digested by bacteria to produce biogas.
3) By-products are used to fertilise crops and gardens.

Anaerobics lesson — keep fit for bacteria...

Fascinating stuff, this biogas. It makes a lot of sense, I suppose, to get energy from rubbish, sewage and pig poop instead of leaving it all to rot naturally — which would mean all that lovely methane just wafting away into the atmosphere. Remember — anaerobic respiration makes biofuels.

Fuels from Microorganisms

Here's more than you could ever have wanted to know about that magic stuff, biogas.

Not All Biogas Generators Are the Same

There are two main types of biogas generator — batch generators and continuous generators.

Batch generators make biogas in small batches. They're manually loaded up with waste, which is left to digest, and the by-products are cleared away at the end of each session.

Continuous generators make biogas all the time. Waste is continuously fed in, and biogas is produced at a steady rate. Continuous generators are more suited to large-scale biogas projects.

The diagram on the right shows a simple biogas generator. Whether it's a continuous or batch generator, it needs to have the following:

1) an inlet for waste material to be put in
2) an outlet for the digested material to be removed through
3) an outlet so that the biogas can be piped to where it is needed

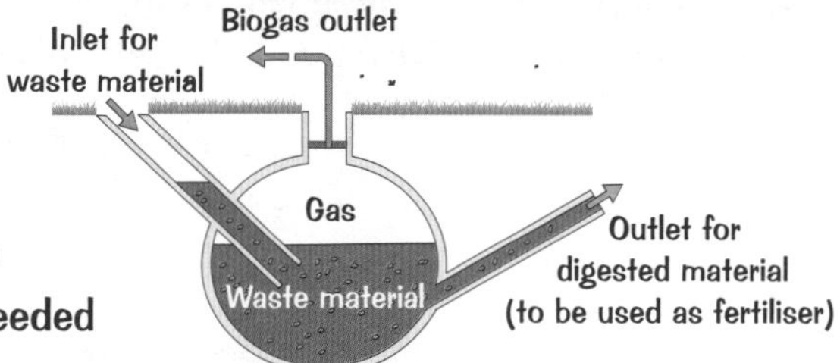

Four Factors to Consider When Designing a Generator:

When biogas generators are being designed, the following factors need to be considered:

COST: Continuous generators are more expensive than batch ones, because waste has to be mechanically pumped in and digested material mechanically removed all the time.

CONVENIENCE: Batch generators are less convenient because they have to be continually loaded, emptied and cleaned.

EFFICIENCY: Gas is produced most quickly at about 35 °C. If the temperature falls below this the gas production will be slower. Generators in some areas will need to be insulated or kept warm, e.g. by solar heaters. The generator shouldn't have any leaks or gas will be lost.

POSITION: The waste will smell during delivery, so generators should be sited away from homes. The generator is also best located fairly close to the waste source.

Using Biofuels Has Economic and Environmental Effects

1) Biofuels are a 'greener' alternative to fossil fuels. The carbon dioxide released into the atmosphere was taken in by plants which lived recently, so they're 'carbon neutral'.
2) The use of biofuels doesn't produce significant amounts of sulfur dioxide or nitrogen oxides, which cause acid rain.
3) Methane is a greenhouse gas and is one of those responsible for global warming. It's given off from untreated waste, which may be kept in farmyards or spread on agricultural land as fertiliser. Burning it as biogas means it's not released into the atmosphere.
4) The raw material is cheap and readily available.
5) The digested material is a better fertiliser than undigested dung — so people can grow more crops.
6) In some developing rural communities women have to spend hours each day collecting wood for fuel. Biogas saves them this drudgery.
7) Biogas generators act as a waste disposal system, getting rid of human and animal waste that'd otherwise lie around, causing disease and polluting water supplies.

CO_2 released

animal waste

Methane changed to CO_2

CO_2 absorbed in photosynthesis

Biogas generator

Don't sit under a cow — unless you want a pat on the head...

Biogas is fantastic. It gets rid of waste, makes a great fertiliser AND provides energy. Biogas isn't new though — before electricity, it was drawn from London's sewer pipes and burned in the street lights.

Using Microorganisms Safely

Microorganisms can be grown in a lab, but they need certain conditions to flourish.
Also, precautions must be taken to stop unwanted microorganisms growing as well.

Microorganisms Are Grown on Agar Jelly in a Petri Dish

1) Microorganisms are grown (cultured) in a "culture medium".
2) They need carbohydrates as an energy source, plus mineral ions, and sometimes supplementary proteins and vitamins.
3) These nutrients are usually added to the agar jelly.
4) Agar jelly can be poured when hot, and sets when cold.
   It's poured into shallow round plastic dishes called Petri dishes.

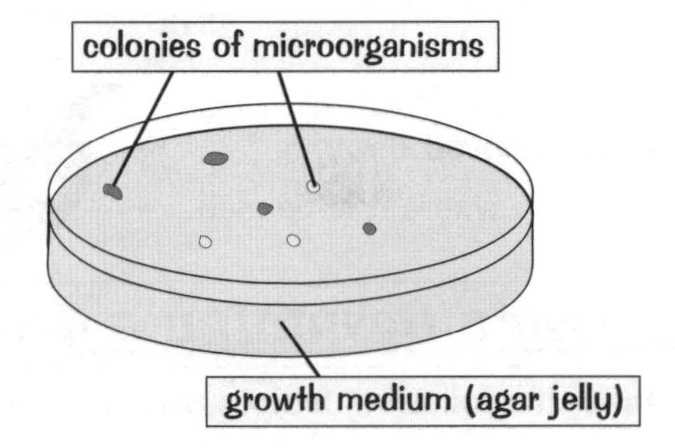

Equipment is Sterilised to Prevent Contamination

1) If equipment isn't sterilised, unwanted microorganisms in the growth medium will grow and contaminate the end product.
2) The unwanted microorganisms might make harmful substances, or cause disease.
3) Petri dishes and the growth medium must be sterilised before use.
4) Inoculating loops (used for transferring microorganisms to the growth medium) are sterilised by passing them through a flame.
5) The Petri dish must have a lid to stop any microorganisms in the air contaminating the culture. The lid should be taped on.

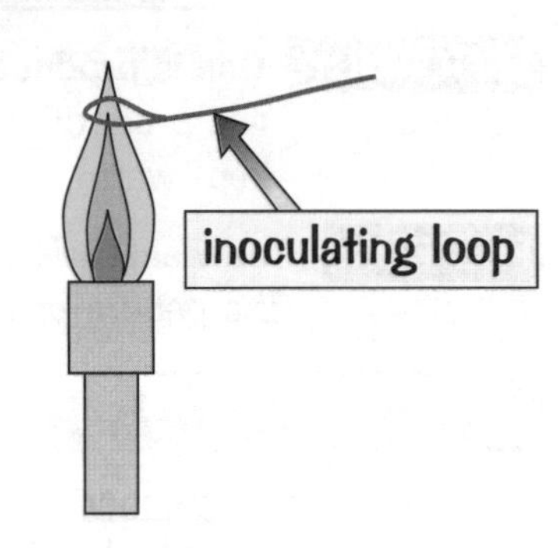

The Temperature Must be Kept Fairly Low in School Labs

Pathogens are microorganisms which cause disease.

In the lab at school, cultures of microorganisms are kept at about 25 °C. Harmful pathogens aren't likely to grow at this temperature.

In industrial conditions, cultures are incubated at higher temperatures so that they can grow a lot faster.

Agar — my favourite jelly flavour after raspberry...

Culture medium = growth medium = liquid or jelly that the microorganisms are grown in. Sorted. Microorganisms might be the perfect pets. You don't have to walk them, they won't get lonely and they cost hardly anything to feed. But whatever you do, do not feed them after midnight.

Revision Summary for Biology 3(ii)

So you think you've learnt these pages on microorganisms, eh... Well, there's only one way to really find out. And you know what that is, I'll bet. It's obvious... I mean, there's a whole load of questions staring you in the face — chances are, it's got to involve those in some way. And sure enough, it does.
Just write down the answers to all these questions. Then go back over the section and see if you got any wrong. If you did, then you need a bit more revision, so go back and have another read of the section and then have another go. It's the best way to make sure you actually know your stuff.

1) What's the difference between spontaneous generation and biogenesis?
2) Describe three experiments that helped develop the theory of biogenesis.
3) What kind of microorganism is used in the manufacture of cheese?
4) In yoghurt-making, what's produced when the bacteria ferment the lactose in milk?
5) What type of microorganisms are yeasts?
6) What does "anaerobic" mean? Write an equation for the anaerobic respiration of glucose by yeast.
7) Write an equation for the aerobic respiration of glucose by yeast.
8)* Some yeast is added to 100 ml of water. Half of this is cooled to 0 °C for 2 hours and the other half is placed in a water bath at 90 °C for 2 hours.
   Each sample is then placed in a flask as shown in the diagram and brought to 40 °C. 20 g of sugar is added to each flask and the amount of carbon dioxide produced by each is measured after an hour.
   The sample that had been cooled to 0 °C produced 18 cm^3 of carbon dioxide. The sample that had been heated did not produce any carbon dioxide.
   What conclusions can you draw from these results? How could you check the reliability of your results?

9) Explain what is meant by "malting".
10) Give four examples of conditions that are controlled inside an industrial fermenter.
11) What microorganism is the main source of mycoprotein?
12) What microorganism is used to make penicillin?
13) What is ethanol used for in some countries (apart from the obvious)?
14) What are the two main components of biogas?
15)* Loompah is a small village. It's very hot in the summer but freezing cold in winter. The villagers keep goats and cows. They also try to grow crops, but the soil isn't very fertile, so it's difficult. The villagers currently rely on wood for fuel for heating and cooking. There's not much of this around, so they spend a lot of time collecting it, preventing them from practising their nail-art.
    a) How suitable do you think biogas would be for this village? Explain the advantages that using biogas would have for the village. What disadvantages or problems might there be?
    b) Loompah starts using biogas and uses the digested material as fertiliser. They compare their crops to the crops grown by the village of Moompah, which uses normal manure as a fertiliser. Loompah's crops are bigger, so they conclude that the digested material is a better fertiliser than manure.
    What do you think of the conclusion they've drawn?
16)* The data below shows the rate of biogas produced by a generator at various temperatures.
    a) Draw a graph showing the temperature against biogas produced. Join the points with a smooth curve.
    b) Using your graph, estimate the optimum temperature for biogas production.
    c) How much biogas would you expect to be produced in 24 hours at 25 °C?

Temperature (°C)	10	20	30	40	50	60
Biogas produced in 1 hour (cm^3)	6	32	54	78	50	18

17) What is a "culture medium"?
18) Why is it important to sterilise laboratory equipment before using it to culture microorganisms?
19) At what temperature are bacterial cultures usually incubated in the class lab? Is this hotter or colder than in industrial labs?

* Answers on page 96

Thinking in Exams

In the old days, it was enough to learn a whole bunch of facts while you were revising and just spew them onto the paper come exam day. If you knew the facts, you had a good chance of doing well, even if you didn't really understand what any of those facts actually meant. But those days are over. Rats.

Remember — You Might Have to Think During the Exam

1) Nowadays, the examiners want you to be able to apply your scientific knowledge to situations you've never seen before. Eeek.
2) The trick is not to panic. They're not expecting you to show Einstein-like levels of scientific insight (not usually, anyway).
3) They're just expecting you to use the science you know in an unfamiliar setting — and usually they'll give you some extra info too that you should use in your answer.

So to give you an idea of what to expect come exam-time, use the new CGP Exam Simulator (below). Read the article, and have a go at the questions. It's guaranteed to be just as much fun as the real thing.

Underlining or making notes of the main bits as you read is a good idea.

1. Blood glucose levels controlled by insulin.
2. Insulin added → liver removes glucose.
3. Not enough insulin → high blood glucose → death?
4. Carbohydrates cause problems for diabetics. So carbohydrates and glucose linked...

All cells need energy to function, and this energy is supplied by glucose carried in the blood. The level of glucose in the blood is controlled by the hormone insulin — if the blood glucose level gets too high, insulin is introduced into the bloodstream, which in turn makes the liver remove glucose.

Diabetes (type I) is where not enough insulin is produced, meaning that a person's blood glucose level can rise to a level that can kill them. The problem can be controlled in two ways:

a) Avoiding foods rich in carbohydrates. It can also be helpful to take exercise after eating carbohydrates.

b) Injecting insulin before meals (especially if high in carbohydrates).

Dave Edwards, a director of InsulinProducts plc, said: "We recommend controlling diabetes via insulin injections for its ease and safety."

Questions:

1. Why can it be helpful for a diabetic to take exercise after eating carbohydrates?
2. Suggest why a diabetic person should make sure they eat sensibly after injecting insulin.
3. Why might some people suspect Dave Edwards of being biased?

Clues — don't read unless you need a bit of a hand...

1. More complex carbohydrates are broken down to make glucose. What would normally happen if lots of glucose is suddenly put into the blood? How would this normally be controlled? And what happens in a diabetic?
2. Think about what insulin causes to happen.
3. What's his job?

Answers

1) Eating carbohydrates puts a lot of glucose into the blood. Exercising can use up extra glucose, which helps stop blood glucose levels getting too high.
2) If they don't, blood glucose levels can drop dangerously low.
3) He's a director of a firm that probably makes insulin — so he'll want to make that sound as good as possible.

Thinking in an exam — it's not like the old days...

It's scary — being expected to think in the exam. But just take your time and think things through. On a lighter note... doctors used to diagnose diabetes by tasting a patient's wee — if the urine tasted sugary, that patient probably had diabetes. How times change...

Answering Experiment Questions (i)

You'll definitely get some questions in the exam about experiments. They can be about any topic under the Sun — but if you learn the basics and throw in a bit of common sense, you'll be fine.

Read the Question Carefully

The question might describe an experiment, e.g. —

Ellie had three different powdered fertilisers: A, B and C.
She investigated which fertiliser was most effective when growing pea plants.
Ellie planted a seed in each of 12 pots. She then added fertiliser A to three pots, fertiliser B to three pots and fertiliser C to three. She did not add any fertiliser to the remaining three pots.

Fertiliser A

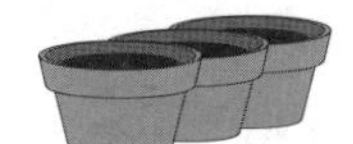
Fertiliser B

Fertiliser C

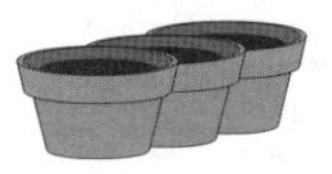
No fertiliser

She watered each pot daily and after three weeks she measured the heights of each seedling.

1. What is the independent variable?
   The type of fertiliser.
2. What is the dependent variable?
   The heights of the seedlings.
3. Give two variables that must be kept the same to make it a fair test.
   1. The amount of fertiliser.
   2. The amount of light.
4. What is the control group in this investigation?
   The group of pots with no fertiliser.

The independent variable is the thing that is changed.

The dependent variable is the thing that's measured.

To make it a fair test, you've got to keep all the other variables the same. Or else you won't know if the only thing affecting the dependent variable is the independent variable.

There are loads of other things that must be kept the same for each pot in this experiment. You could also have put temperature or the type of soil, etc.

It's easy to keep the variables the same in this experiment as it's in a laboratory. But it can sometimes be trickier. For example, if the seeds were growing in fields, it'd be hard to make sure that they all had exactly the same soil conditions, and got the same amount of water and light, etc.

It's even harder to make investigations involving people fair.

If, say, the effect of a person's age on their blood pressure was being investigated, there'd be loads of other variables to consider — weight, diet and whether someone's a smoker could make a big difference to their blood pressure.

To make it a fairer test, it would be better if just nonsmokers with a similar weight and diet were used.

A control group isn't really part of the experiment, but it's kept in the same conditions as the rest of the experiment. You can compare changes in the experiment with those that happened to the control group, and see if the changes might have happened anyway. Control groups make results more meaningful.
In this experiment the seeds might grow better without any fertiliser — with a control group you can check for this.

Control groups are used when testing drugs. People can feel better just because they've been given a drug that they believe will work. To rule this out, researchers give one group of patients dummy pills (called placebos) — but they don't tell them that their pills aren't the real thing.
This is the control group. By doing this, they can tell if the real drug is actually working.

Answering Experiment Questions (ii)

5. Why was each type of fertiliser added to three pots, instead of just one?

   To check for anomalous results and make the results more reliable.

Sometimes unusual results are produced — repeating an experiment gives you a better idea what the correct result should be.

6. The table below shows the heights of the seedlings in each pot.

	First pot	Second pot	Third pot	Mean
Fertiliser A	4.4 cm	5.2 cm	4.2 cm	
Fertiliser B	8.3 cm	7.9 cm	8.7 cm	8.3 cm
Fertiliser C	6.7 cm	5.7 cm	0 cm	6.2 cm
No fertiliser	2.4 cm	1.9 cm	2.6 cm	2.3 cm

   a) Calculate the mean height of the seedlings grown with fertiliser A.

   Mean = (4.4 + 5.2 + 4.2) ÷ 3 = 4.6 cm

   b) What is the range of the heights of the seedlings grown with fertiliser A?

   5.2 − 4.2 = 1.0 cm

When an experiment is repeated, the results will usually be slightly different each time. The mean (or average) of the measurements is usually used to represent the values.

The more times the experiment is repeated the more reliable the average will be.

To find the mean:

ADD TOGETHER all the data values and DIVIDE by the total number of values in the sample.

The range is how far the data spreads. You just work out the difference between the highest and lowest numbers.

7. One of the results in the table is anomalous. Circle the result and suggest why it may have occurred.

   The seed may have had something genetically wrong with it.

If one of the results doesn't seem to fit in, it's called an anomalous result. You should usually ignore an anomalous result. It's been ignored when the mean was worked out.

This is a random error — it only happens occasionally.

If the same mistake is made every time, it's a systematic error, e.g. if you measured from the very end of your ruler instead of from the 0 cm mark every time, meaning all your measurements would be a bit small.

8. What conclusion can you draw from these results?

   Fertiliser B makes pea plants grow taller over the first three weeks than fertilisers A or C, given daily watering.

Be careful that your conclusions match the data you've got, and don't go any further.

You can't say that fertiliser B will always be better than fertilisers A or C, because:

- The results may be totally different with another type of plant.
- After four weeks, the plants grown with fertiliser B may all drop dead, while the others keep growing. Etc.

Mistakes happen...

NASA made a bit of a boo-boo once. They muddled up measurements in pounds and newtons and caused the Mars Climate Orbiter to burn up in the Martian atmosphere. Ooops. It just goes to show that anyone can make a mistake, even a bunch of brainy boffins. So always double-check everything.

Answering Experiment Questions (iii)

Use Sensible Measurements for Your Variables

Pu-lin did an experiment to see how the mass of a potato changed depending on the sugar solution it was in.

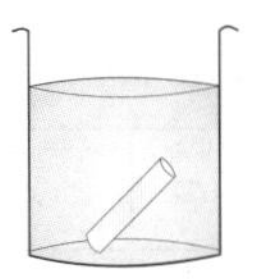

She started off by making potato tubes 5 cm in length, 1 cm in diameter and 2.0 g in mass. She then filled a beaker with 500 ml of pure water and placed a potato tube in it for 30 minutes. She repeated the experiment with different amounts of sugar dissolved in the water. For each potato tube, she measured the new mass. She did the experiment using Charlotte, Desiree, King Edward and Maris Piper potatoes.

Before she started, she did a trial run, which showed that most of the potato tubes shrunk to a minimum of 1 g (in a really strong sugar solution) or grew to a maximum of 3 g (in pure water).

1. What kind of variable was the list of potatoes?
   - **A** A continuous variable ☐
   - **B** A categoric variable ☑
   - **C** An ordered variable ☐
   - **D** A discrete variable ☐

2. Pu-lin should add sugar in intervals of...
   - **A** a pinch ☐
   - **B** a teaspoon ☑
   - **C** a cupful ☐
   - **D** a bucketful ☐

3. The balance used to find the mass of the potato should be capable of measuring...
   - **A** to the nearest 0.01 gram ☑
   - **B** to the nearest 0.1 gram ☐
   - **C** to the nearest gram ☐
   - **D** to the nearest 10 grams ☐

Categoric variables are variables that can't be related to size or quantity — they're types of things. E.g. names of potatoes or types of fertiliser.

Continuous data is numerical data that can have any value within a range — e.g. length, volume, temperature and time.

Note: You can't measure the exact value of continuous data. Say you measure a height as 5.6 cm to the nearest mm. It's not exact — you get a more precise value if you measure to the nearest 0.1 mm or 0.01 mm, etc.

Ordered variables are things like small, medium and large lumps, or warm, very warm and hot.

Discrete data is the type that can be counted in chunks, where there's no in-between value. E.g. number of people is discrete, not continuous, because you can't have half a person.

It's important to use sensible values for variables. It's no good using loads of sugar or really weedy amounts like a pinch at a time cos you'd be there forever and the results wouldn't show any significant difference. (You'd get different amounts of sugar in each pinch anyway.)

A balance measuring only to the nearest gram, or bigger, would not be sensitive enough — the changes in mass are likely to be quite small, so you'd need to measure to the nearest 0.01 gram to get the most precise results.

The sensitivity of an instrument is the smallest change it can detect, e.g. some balances measure to the nearest gram, but really sensitive ones measure to the nearest hundredth of a gram. For measuring tiny changes — like from 2.00 g to 1.92 g — the more sensitive balance is needed.

You also have to think about the precision and accuracy of your results. Precise results are ones taken with sensitive instruments, e.g. volume measured with a burette will be more precise than volume measured with a 100 ml beaker. Really accurate results are those that are really close to the true answer. It's possible for results to be precise but not very accurate, e.g. a fancy piece of lab equipment might give results that are precise, but if it's not calibrated properly those results won't be accurate.

I take my tea milky with two bucketfuls of sugar... mmm...

Accuracy, precision and sensitivity are difficult things to get your head around — a sensitive piece of equipment is likely to give precise results (but not necessarily very accurate results). If the equipment is used properly and calibrated well then the results are more likely to be accurate...

Answering Experiment Questions (iv)

Once you've collected all your data together, you need to analyse it to find any relationships between the variables. The easiest way to do this is to draw a graph, then describe what you see...

Graphs Are Used to Show Relationships

These are the results Pu-lin obtained with the King Edward potato.

Number of teaspoons of sugar	0	2	4	6	8	10	12	14	16	18	20
Mass of potato tube (g)	2.50	2.40	2.23	2.10	2.02	1.76	1.66	1.25	1.47	1.3	1.15

4. a) Nine of the points are plotted below.
   Plot the remaining **two** points on the graph.

To plot the points, use a sharp pencil and make a neat little cross.

nice clear mark

   b) Draw a straight line of best fit for the points.

A line of best fit is drawn so that it's easy to see the relationship between the variables. You can then use it to estimate other values.

When drawing a line of best fit, try to draw the line through or as near to as many points as possible, ignoring any anomalous results.

Scattergram to show the mass of a King Edward potato tube in different sugar solutions

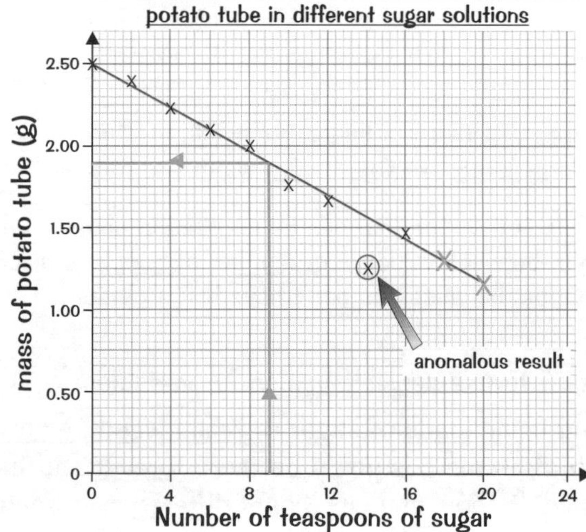

This is a scattergram — they're used to see if two variables are related.

This graph shows a negative correlation between the variables. This is where one variable increases as the other one decreases.

The other correlations you could get are:

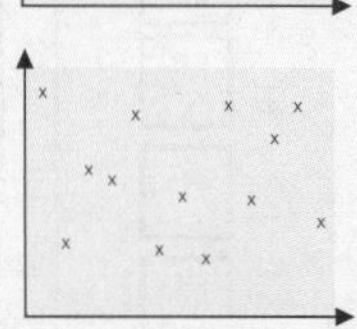

Positive correlation — this is where as one variable increases so does the other one.

No correlation — this is where there's no obvious relationship between the variables.

5. Estimate the mass of the potato tube if you added nine teaspoons of sugar.

   Estimate of mass = 1.90 g (see graph)

6. What can you conclude from these results?

   There is a negative correlation between the number of teaspoons of sugar and the mass of potato tube. Each additional teaspoon causes the potato tube to lose mass.

In lab-based experiments like this one, you can say that one variable causes the other one to change. The extra sugar causes the potato to lose mass. You can say this because everything else has stayed the same — nothing else could be causing the change.

There's a positive correlation between revising and good marks...

...really, it's true. Other ways to improve your marks are to practise plotting graphs, and learning how to read them properly — make sure you're reading off the right axis for a start, and don't worry about drawing lines on the graph if it helps you to read it. Always double-check your answer... just in case.

Answering Experiment Questions (v)

Not all experiments can be carefully controlled in a laboratory. Some have to be done in the real world. Unfortunately, this creates complications of its own.

Relationships Do NOT Always Tell Us the Cause

Melanomas are a dangerous form of skin cancer. It's thought that UV damage may increase the risk of getting skin cancer later in life, so people are advised to avoid being in direct sunlight for long periods at a time.

The graph shows the number of new cases of melanoma found per year in people who spend at least 5 hours of each working day exposed to direct sunlight.

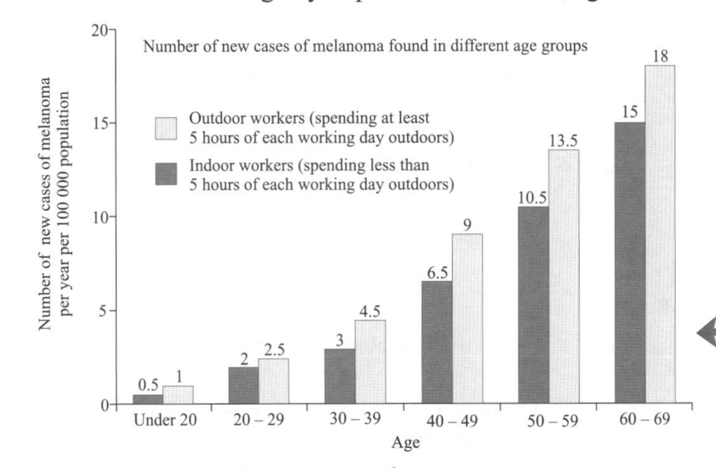

In large studies done outside a lab it's really difficult to keep all the variables the same and to make sure the control group are kept in the same conditions.

In this study the control group are people who work indoors.

This is a bar chart. It contains a key to tell you which colour bars relate to which group.

1. The graph is representative of a country that has 500 000 people aged between 40 and 49 who work outdoors.

   Using the graph, estimate how many of these people will get melanomas in a given year.

   9 × 5 = 45 people

2. What conclusion can you draw from these results?

   Melanomas are found more frequently in people who spend at least 5 hours of each working day in direct sun than in people who work indoors.

3. Suggest how the data may have been collected.

   e.g. from medical records

Here they're asking you to estimate the number of people out of 500 000 — the graph tells you the number of people in 100 000. Don't get caught out — read the question really carefully.

When describing the data and drawing conclusions it's really important that you don't say that working out in the sun causes melanomas. The graph only shows that there's a positive correlation between the two.

In studies like this where you're unable to control everything, it's possible a third variable is causing the relationship. For example, many people who work outside are farm labourers or builders. Pesticides and other toxic chemicals that you might be exposed to in these professions could cause the increased rate of melanomas.

Use your common sense to think of a sensible answer.

Try to suggest a method to get reliable results. For example, it's very unlikely that the data would have been collected by a telephone survey or an internet search.

Who wants to live in the real world anyway... (I have my very own special one)

It's really difficult to prove what causes what in science, especially with all the things you've got to control. The experiments are usually done in a lab first so that you can control as much as possible. Then they're done in the real world to see if the same thing happens, and to find any unexpected results.

Answers

Page 12

8) a) A, b) B

14) b) Day 1 = A relaxing day at home,
Day 2 = Sitting on Blackpool beach in January,
Day 3 = Running a marathon
c) You'd need to consume a lot more water on day 3 than on days 1 or 2.

Page 23

2) professional runner, builder, waitress, secretary

12) a) 6 pm, b) 8 pm, c) No

16) a) Baka-Lite Bread has less of everything per slice, but more per 100 g. The Baka-Lite Bread slices are just a lot smaller than Standard Bread slices.
b) Standard Bread is healthier. It has less salt and saturated fat than Baka-Lite Bread does. Standard Bread also has more fibre.

Page 33

1) a) Big penguins have a smaller surface area to volume ratio than small penguins. So big penguins will lose a smaller proportion of their body heat and so can survive in colder temperatures.
b) E.g. webbed feet for swimming to catch fish, fluffy underneath feathers trap air for insulation, a layer of fat under skin for insulation, oily feathers help waterproofing.

2) The kangaroo rat will probably produce a very small amount of concentrated urine. It will probably sweat very little. It will probably have a large surface area to maximise heat loss.

7) a) 1875 approx.
b) It started to fall, this may be because grey squirrels outcompeted red squirrels for resources.
c) Food, space, shelter, mates, water, nesting sites.

22) b) The theory is supported by fossil evidence.

Page 38

10) The fact that one glacier is melting doesn't mean that all glaciers are melting.
One glacier melting doesn't mean that the average global temperature is rising.
You'd need to collect a lot more data from around the whole world over a long period of time.

12) a) As the world population increases the number of extinct species increases.
b) More humans means more animals hunted and more habitats are destroyed to make way for farming, living etc.

15) a) 25, b) 2002, c) 0.5 tonnes, d) 2003,
e) Conserves finite resources such as metals; reduces landfill; uses less energy, meaning less CO_2 is released.

Page 62

5) a) 1.65, b) stomach.

Page 72

16) 50%

23) BB and bb.

Page 82

20) a) The athlete's body will make more red blood cells to compensate for the lower oxygen levels at night.
The increased number of red blood cells will allow the athlete to get more oxygen to his or her muscle cells for respiration, meaning more energy is released.
This may improve their performance.
b) Live/sleep at a high altitude, e.g. up a mountain.
Inject red blood cells.

22) a)

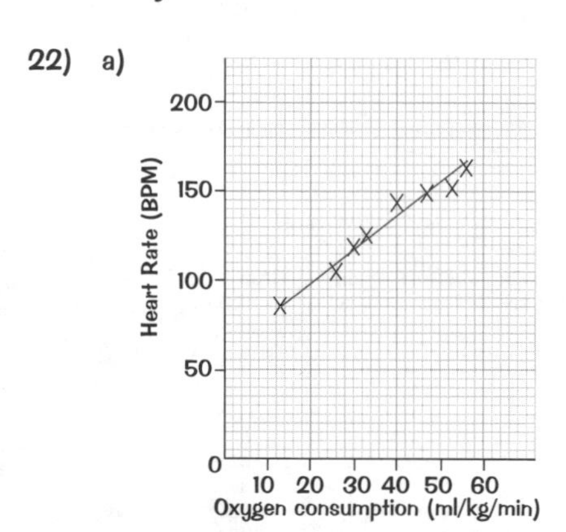

b) It shows that oxygen consumption increases as heart rate increases.
c) When the athlete exercises their muscles consume more oxygen (because they need more energy to contract faster). The heart rate needs to increase in order to supply more oxygen.

Page 89

8) Heating yeast to 90 °C for 2 hours kills it.
Keeping yeast at 0 °C doesn't kill it.
You could repeat your experiment to check if you get the same results.

15) a) Biogas is suitable because waste from the goats and cows can be used in the biogas generator.
Advantages: villagers won't have to spend time collecting wood, digested material could be used to fertilise soil, and waste would be disposed of, reducing disease.
Disadvantages: biogas production slows down in cold conditions, so they might need an alternative fuel source in winter.
b) Their conclusion isn't valid. Possible reasons: the amounts spread on the ground might have been different, the weather in the two places might have been different, the species of crop might have been different etc.

16) a)

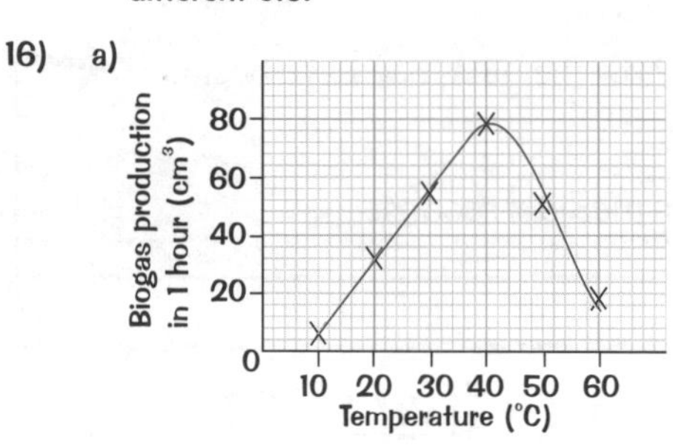

b) 40 °C
c) 44 cm^3 in 1 hour so it's 44 x 24 = 1056 cm^3 in 24 hours.